MATLAB®

An Introduction with Applications

Third Edition

Amos Gilat

Department of Mechanical Engineering
The Ohio State University

BICENTENNIAL
1807
WILEY
2007
BICENTENNIAL

JOHN WILEY & SONS, INC.

EXECUTIVE PUBLISHER   Don Fowley
ASSOCIATE PUBLISHER   Dan Sayre
ACQUISITIONS EDITOR   Michael McDonald
EDITORIAL ASSISTANT   Rachael Leblond
SENIOR PRODUCTION EDITOR   Ken Santor
COVER DESIGNER   Michael St. Martine

This book was set in Adobe Framemaker® by the author and printed and bound by Malloy Inc. The cover was printed by Malloy Inc.

This book is printed on acid free paper. ∞

ISBN 978-0-470-10877-2

Printed in the United States of America

10 9 8 7 6 5 4 3 2 1

N

A

w   ns

1807
WILEY
2007

BICENTENNIAL
BICENTENNIAL
BICENTENNIAL
BICENTENNIAL

THE WILEY BICENTENNIAL—KNOWLEDGE FOR GENERATIONS

$\mathcal{E}$ach generation has its unique needs and aspirations. When Charles Wiley first opened his small printing shop in lower Manhattan in 1807, it was a generation of boundless potential searching for an identity. And we were there, helping to define a new American literary tradition. Over half a century later, in the midst of the Second Industrial Revolution, it was a generation focused on building the future. Once again, we were there, supplying the critical scientific, technical, and engineering knowledge that helped frame the world. Throughout the 20th Century, and into the new millennium, nations began to reach out beyond their own borders and a new international community was born. Wiley was there, expanding its operations around the world to enable a global exchange of ideas, opinions, and know-how.

For 200 years, Wiley has been an integral part of each generation's journey, enabling the flow of information and understanding necessary to meet their needs and fulfill their aspirations. Today, bold new technologies are changing the way we live and learn. Wiley will be there, providing you the must-have knowledge you need to imagine new worlds, new possibilities, and new opportunities.

Generations come and go, but you can always count on Wiley to provide you the knowledge you need, when and where you need it!

WILLIAM J. PESCE
PRESIDENT AND CHIEF EXECUTIVE OFFICER

PETER BOOTH WILEY
CHAIRMAN OF THE BOARD

Preface

MATLAB® is a very popular language for technical computing used by students, engineers, and scientists in universities, research institutes, and industries all over the world. The software is popular because it is powerful and easy to use. For university freshmen in it can be thought of as the next tool to use after the graphic calculator in high school.

This book was written following several years of teaching the software to freshmen in an introductory engineering course. The objective was to write a book that teaches the software in a friendly, non-intimidating fashion. Therefore, the book is written in simple and direct language. In many places bullets, rather than lengthy text, are used to list facts and details that are related to a specific topic. The book includes numerous sample problems in mathematics, science, and engineering that are similar to problems encountered by new users of MATLAB.

This third edition of the book is updated for MATLAB 7.5 (Release 2007b). Other modifications/changes to this edition are: script files are introduced in Chapter 1 (this allows students to use script files for solving problems in Chapters 2 and 3), new coverage of the Workspace Window, the save and load commands, plotting figures with error bars, and instructions for using several Figure Windows at the same time. Chapter 6 was revised to include coverage on anonymous functions, function functions, function handles, subfunctions and nested functions. In addition, the end of chapter problems have been revised. There are many new problems (more than half), and the problems cover a wider range of topics.

I would like to thank several of my colleagues at The Ohio State University. Professors Richard Freuler, Mark Walter, and Walter Lampert, and Dr. Mike Parke read sections of the book and suggested modifications. I also appreciate the involvement and support of Professors Robert Gustafson and John Demel and Dr. John Merrill from the First-Year Engineering Program at The Ohio State University. Special thanks go to Professor Mike Lichtensteiger (OSU), and my daughter Tal Gilat (Marquette University), who carefully reviewed the first edition of the book and provided valuable comments and criticisms. Professor Brian Harper (OSU) has made a significant contribution to the new end of chapter problems in the present edition.

I would like to express my appreciation to all those who have reviewed the first edition of the text at its various stages of development, including Betty Barr, University of Houston; Andrei G. Chakhovskoi, University of California, Davis; Roger King, University of Toledo; Richard Kwor, University of Colorado at Colorado Springs; Larry Lagerstrom, University of California, Davis; Yueh-Jaw Lin, University of Akron; H. David Sheets, Canisius College; Geb Thomas, University of Iowa; Brian Vick, Virginia Polytechnic Institute and State University; Jay Weitzen, University of Massachusetts, Lowell; and Jane Patterson Fife, The Ohio State University. In addition, I would like to acknowledge Gladys Soto, Ken Santor, and Rachael Leblond, all from John Wiley & Sons, who supported the production of the third edition.

I hope that the book will be useful and will help the users of MATLAB to enjoy the software.

Amos Gilat
Columbus, Ohio
November, 2007
gilat.1@osu.edu

To my parents Schoschana and Haim Gelbwacks

Contents

Introduction

MATLAB is a powerful language for technical computing. The name MATLAB stands for MATrix LABoratory, because its basic data element is a matrix (array). MATLAB can be used for math computations, modeling and simulations, data analysis and processing, visualization and graphics, and algorithm development.

MATLAB is widely used in universities and colleges in introductory and advanced courses in mathematics, science, and especially in engineering. In industry the software is used in research, development and design. The standard MATLAB program has tools (functions) that can be used to solve common problems. In addition, MATLAB has optional toolboxes that are collections of specialized programs designed to solve specific types of problems. Examples include toolboxes for signal processing, symbolic calculations, and control systems.

Until recently, most of the users of MATLAB have been people who had previous knowledge of programming languages such as FORTRAN or C, and switched to MATLAB as the software became popular. Consequently, the majority of the literature that has been written about MATLAB assumes that the reader has knowledge of computer programming. Books about MATLAB often address advanced topics or applications that are specialized to a particular field. In the last few years, however, MATLAB is being introduced to college students as the first (and sometimes the only) computer program they learn. For these students there is a need for a book that teaches MATLAB assuming no prior experience in computer programming.

The Purpose of This Book

MATLAB: An Introduction with Applications is intended for students who are using MATLAB for the first time and have little or no experience in computer programming. It can be used as a textbook in freshmen engineering courses, or workshops where MATLAB is being taught. The book can also serve as a reference in more advanced science and engineering courses when MATLAB is used as a tool for solving problems. It also can be used for self study of MATLAB by students and practicing engineers. In addition, the book can be a supplement or a secondary book in courses where MATLAB is used, but the instructor does not have the time to cover it extensively.

Topics Covered

MATLAB is a huge program, and therefore it is impossible to cover all of it in one book. This book focuses primarily on the foundations of MATLAB. It is believed

1

that once these foundations are well understood, the student will be able to learn advanced topics easily by using the information in the Help menu.

The order in which the topics are presented in this book was chosen carefully, based on several years of experience in teaching MATLAB in an introductory engineering course. The topics are presented in an order that allows the student to follow the book chapter after chapter. Every topic is presented completely in one place and then is used in the following chapters.

The first chapter describes the basic structure and features of MATLAB and how to use the program for simple arithmetic operations with scalars as with a calculator. Script files are introduced at the end of the chapter. They allow the student to write, save, and execute simple MATLAB programs. The next two chapters are devoted to the topic of arrays. MATLAB's basic data element is an array that does not require dimensioning. This concept, which makes MATLAB a very powerful program, can be a little difficult to grasp for students who have only limited knowledge and experience with linear algebra and vector analysis. The book is written so that the concept of arrays is introduced gradually and then explained in extensive detail. Chapter 2 describes how to create arrays, and Chapter 3 covers mathematical operations with arrays.

Following the basics, more advanced topics that are related to script files and input and output of data are presented in Chapter 4. This is followed by two-dimensional plotting that is covered in Chapter 5. User-defined functions and function files are covered next in Chapter 6. The coverage of function files is intentionally separated from the subject of script files. This has been proven to be easier to understand by students who are not familiar with similar concepts from other computer programs. Programming with MATLAB is covered in Chapter 7, which includes flow control with conditional statements and loops.

The next three chapters cover more advanced topics. Chapter 8 describes how MATLAB can be used for carrying out calculations with polynomials, and how to use MATLAB for curve fitting and interpolation. Plotting three-dimensional plots, which is an extension of the chapter on two-dimensional plots, is covered in Chapter 9. Chapter 10 covers applications of MATLAB for numerical analysis. It includes solving nonlinear equations, finding a minimum or a maximum of a function, numerical integration, and solution of first order ordinary differential equations. Chapter 11 covers in great detail how to use MATLAB in symbolic operations.

The Framework of a Typical Chapter

In every chapter the topics are introduced gradually in an order that makes the concepts easy to understand. The use of MATLAB is demonstrated extensively within the text and by examples. Some of the longer examples in Chapters 1–3 are titled as tutorials. Every use of MATLAB is printed in the book with a different font and with gray background. Additional explanations appear in boxed text with white background. The idea is that the reader will execute these demonstrations and tutorials in order to gain experience in using MATLAB. In addition, every chapter includes formal sample problems that are examples of applications of MATLAB for solving problems in math, science, and engineering. Each example includes a problem statement and a detailed solution. Some sample problems are presented in the middle of the chapter. All of the chapters (except Chapter 2) have a section at the end with several sample problems of applications. It should be pointed out that problems with MATLAB can be solved in many different ways. The solutions of the sample problems are written such that they are easy to follow. This means that in many cases the problem can be solved by writing a shorter, or sometimes "trickier," program. The students are encouraged to try to write their own solutions and compare the end results. At the end of each chapter there is a set of homework problems. They include general problems from math and science and problems from different disciplines of engineering.

Symbolic Calculations

MATLAB is essentially a software for numerical calculations. Symbolic math operations, however, can be executed if the Symbolic Math toolbox is installed. The Symbolic Math toolbox is included in the student version of the software and can be added to the standard program.

Software and Hardware

The MATLAB program, like most other software, is continually being developed and new versions are released frequently. This book covers MATLAB, Version 7.5, Release 2007b. It should be emphasized, however, that the book covers the basics of MATLAB, which do not change much from version to version. The book covers the use of MATLAB on computers that use the Windows operating system. Everything is essentially the same when MATLAB is used on other machines. The user is referred to the documentation of MATLAB for details on using MATLAB on other operating systems. It is assumed that the software is installed on the computer, and the user has basic knowledge of operating the computer.

The Order of Topics in the Book

It is probably impossible to write a textbook where all the subjects are presented in an order that is suitable for everyone. The order of topics in this book is such that the fundamentals of MATLAB are covered first (arrays and array operations), and, as mentioned before, every topic is covered completely in one location, which makes the book easy to use as a reference.

Chapter 1

Starting with MATLAB

This chapter begins by describing the characteristics and purposes of the different windows in MATLAB. Next, the Command Window is introduced in detail. Chapter 1 shows how to use MATLAB for arithmetic operations with scalars, similar to the way that a calculator is used. This includes the use of elementary math functions with scalars. The chapter then shows how to define scalar variables (the assignment operator) and how to use these variables in arithmetic calculations. The last section in the chapter introduces script files. It shows how to write, save, and execute simple MATLAB programs.

1.1 STARTING MATLAB, MATLAB WINDOWS

It is assumed that the software is installed on the computer, and that the user can start the program. Once the program starts, the MATLAB desktop window that opens, shown in Figure 1-1, contains three smaller windows which are the Command Window, the Current Directory Window, and the Command History Window. This is the default view of MATLAB. These are three of the various windows in MATLAB. A list of the several windows and their purpose is given in Table 1-1. The **Start** button on the lower left side can be used to access MATLAB tools and features.

Four of the windows, the Command Window, the Figure Window, the Editor Window, and the Help Window, which are used extensively throughout the book, are briefly described on the following pages. More detailed descriptions are included in the chapters where they are used. The Command History Window, Current Directory Window, and the Workspace Window are described in Sections 1.2, 1.8.4, and 4.1, respectively.

Command Window: The Command Window is MATLAB's main window, and opens when MATLAB is started. It is convenient to have the Command Window as the only visible window, and this can be done by either closing all the other windows (click on the **x** at the top right-hand side of the window you want to close), or first selecting on the **Desktop Layout** in the **Desktop** menu, and then **Command Window Only** from the submenu that opens. How to work in the Command Window is described in detail in Section 1.2.

Figure 1-1: **The default view of MATLAB desktop.**

Table 1-1: **MATLAB Windows**

Window	Purpose
Command Window	Main window, enters variables, runs programs.
Figure Window	Contains output from graphic commands.
Editor Window	Creates and debugs script and function files.
Help Window	Provides help information.
Launch Pad Window	Provides access to tools, demos, and documentation.
Command History Window	Logs commands entered in the Command Window.
Workspace Window	Provides information about the variables that are used.
Current Directory Window	Shows the files in the current directory.

Figure Window: The Figure Window opens automatically when graphics commands are executed, and contains graphs created by these commands. An example of a Figure Window is shown in Figure 1-2. A more detailed description of this window is given in Chapter 5.

Figure 1-2: **Example of a Figure Window.**

Editor Window: The Editor Window is used for writing and editing programs. This window is opened from the **File** menu in the Command Window. An example of an Editor Window is shown in Figure 1-3. More details on the Editor Window are given in Section 1.8.2 where it is used for creating script files, and in Chapter 6 where it is used to create function files.

Figure 1-3: **Example of an Editor Window.**

Help Window: The Help Window contains help information. This window can be opened from the **Help** menu in the toolbar of any MATLAB window. The Help Window is interactive and can be used to obtain information on any feature of MATLAB. Figure 1-4 shows an open Help Window.

Figure 1-4: The Help Window.

When MATLAB is started for the first time the screen looks like that shown in Figure 1-1 on page 6. For most beginners it is probably more convenient to close all the windows except the Command Window. (Each of the windows can be closed by clicking on the ⊠ button.) The closed windows can be reopened by selecting them from the **Desktop** menu. The windows shown in Figure 1-1 can be displayed by first selecting **Desktop Layout** in the **Desktop** menu and then **Default** from the submenu. The various windows in Figure 1-1 are docked to the desktop. The windows can be undocked (become a separate independent window) by clicking on the ⬈ button on the upper right-hand corner. An independent window can be docked back by clicking on the ⬎ button.

1.2 *WORKING IN THE COMMAND WINDOW*

The Command Window is MATLAB's main window, and can be used for executing commands, opening other windows, running programs written by the user, and managing the software. An example of the Command Window, with several simple commands that will be explained later in this chapter, is shown in Figure 1-5.

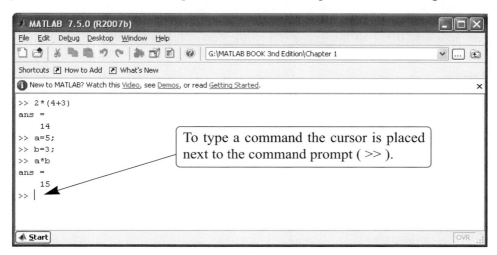

Figure 1-5: The Command Window.

Notes for working in the Command Window:

- To type a command the cursor must be placed next to the command prompt (>>).

- Once a command is typed and the **Enter** key is pressed, the command is executed. However, only the last command is executed. Everything executed previously is unchanged.

- Several commands can be typed in the same line. This is done by typing a comma between the commands. When the **Enter** key is pressed the commands are executed in order from left to right.

- It is not possible to go back to a previous line in the Command Window, make a correction, and then re-execute the command.

- A previously typed command can be recalled to the command prompt with the up-arrow key (↑). When the command is displayed at the command prompt, it can be modified if needed and executed. The down-arrow key (↓) can be used to move down the previously typed commands.

- If a command is too long to fit in one line, it can be continued to the next line by typing three periods … (called an ellipsis) and pressing the **Enter** key. The continuation of the command is then typed in the new line. The command can continue line after line up to a total of 4096 characters.

The semicolon (;):

When a command is typed in the Command Window and the **Enter** key is pressed, the command is executed. Any output that the command generates is displayed in the Command Window. If a semicolon (;) is typed at the end of a command the output of the command is not displayed. Typing a semicolon is useful when the result is obvious or known, or when the output is very large.

 If several commands are typed in the same line, the output from any of the commands will not be displayed if a semicolon is typed between the commands instead of a comma.

Typing %:

When the symbol % (percent symbol) is typed in the beginning of a line, the line is designated as a comment. This means that when the **Enter** key is pressed the line is not executed. The % character followed by text (comment) can also be typed after a command (in the same line). This has no effect on the execution of the command.

 Usually there is no need for comments in the Command Window. Comments, however, are frequently used in programs to add descriptions, or to explain the program (see Chapters 4 and 6).

The `clc` command:

The `clc` command (type `clc` and press **Enter**) clears the Command Window. After working in the Command Window for a while, the display may be very long. Once the `clc` command is executed a clear window is displayed. The command does not change anything that was done before. For example, if some variables were defined previously (see Section 1.6), they still exist and can be used. The up-arrow key can also be used to recall commands that were typed before.

The Command History Window:

The Command History Window lists the commands that have been previously entered in the Command Window. This includes commands from previous sessions. A command in the Command History Window can be used again in the Command Window. By double-clicking on the command, the command is reentered in the Command Window and executed. It is also possible to drag the command to the Command Window, make changes if needed, and then execute it. The list in the Command History Window can be cleared by selecting the lines to be deleted and then selecting **Delete Selection** from the **Edit** menu (or right-click the mouse when the lines are selected and then choose **Delete Selection** in the menu that opens).

1.3 ARITHMETIC OPERATIONS WITH SCALARS

In this chapter we discuss only arithmetic operations with scalars, which are numbers. As will be explained later in the chapter, numbers can be used in arithmetic calculations directly (as with a calculator), or they can be assigned to variables, which can subsequently be used in calculations. The symbols of arithmetic operations are:

Operation	Symbol	Example
Addition	+	$5 + 3$
Subtraction	−	$5 - 3$
Multiplication	*	$5 * 3$
Right division	/	$5 / 3$
Left division	\	$5 \backslash 3 = 3 / 5$
Exponentiation	^	$5 \wedge 3$ (means $5^3 = 125$)

It should be pointed out here that all the symbols except the left division are the same as in most calculators. For scalars, the left division is the inverse of the right division. The left division, however, is mostly used for operations with arrays, which are discussed in Chapter 3.

1.3.1 Order of Precedence

MATLAB executes the calculations according to the order of precedence displayed below. This order is the same as used in most calculators.

Precedence	Mathematical Operation
First	Parentheses. For nested parentheses, the innermost are executed first.
Second	Exponentiation.
Third	Multiplication, division (equal precedence).
Fourth	Addition and subtraction.

In an expression that has several operations, higher-precedence operations are executed before lower-precedence operations. If two or more operations have the same precedence, the expression is executed from left to right. As illustrated in the next section, parentheses can be used to change the order of calculations.

1.3.2 Using MATLAB as a Calculator

The simplest way to use MATLAB is as a calculator. This is done in the Command Window by typing a mathematical expression and pressing the **Enter** key. MATLAB calculates the expression and responds by displaying ans = and the numerical result of the expression in the next line. This is demonstrated in Tutorial 1-1.

Tutorial 1-1: Using MATLAB as a calculator.

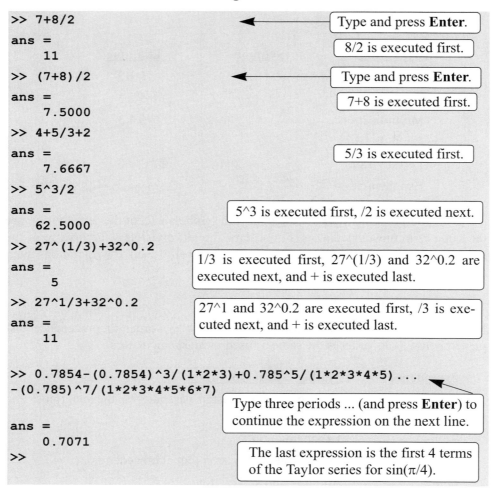

```
>> 7+8/2                                      ◄──────  Type and press Enter.

ans =
    11                                                 8/2 is executed first.

>> (7+8)/2                                    ◄──────  Type and press Enter.

ans =
    7.5000                                             7+8 is executed first.

>> 4+5/3+2

ans =
    7.6667                                             5/3 is executed first.

>> 5^3/2

ans =
    62.5000                            5^3 is executed first, /2 is executed next.

>> 27^(1/3)+32^0.2
                                       1/3 is executed first, 27^(1/3) and 32^0.2 are
ans =                                  executed next, and + is executed last.
    5

>> 27^1/3+32^0.2                       27^1 and 32^0.2 are executed first, /3 is exe-
ans =                                  cuted next, and + is executed last.
    11


>> 0.7854-(0.7854)^3/(1*2*3)+0.785^5/(1*2*3*4*5)...
-(0.785)^7/(1*2*3*4*5*6*7)
                                       Type three periods ... (and press Enter) to
ans =                                  continue the expression on the next line.
    0.7071
>>                                     The last expression is the first 4 terms
                                       of the Taylor series for sin(π/4).
```

1.4 DISPLAY FORMATS

The user can control the format in which MATLAB displays output on the screen. In Tutorial 1-1, the output format is fixed-point with 4 decimal digits (called short), which is the default format for numerical values. The format can be changed with the format command. Once the format command is entered, all the output that follows is displayed in the specified format. Several of the available formats are listed and described in Table 1-2.

MATLAB has several other formats for displaying numbers. Details of these formats can be obtained by typing help format in the Command Window. The format in which numbers are displayed does not affect how MATLAB computes and saves numbers.

Table 1-2: Display formats

Command	Description	Example
format short	Fixed-point with 4 decimal digits for: $0.001 \le number \le 1000$ Otherwise display format short e.	>> 290/7 ans = 41.4286
format long	Fixed-point with 14 decimal digits for: $0.001 \le number \le 100$ Otherwise display format long e.	>> 290/7 ans = 41.42857142857143
format short e	Scientific notation with 4 decimal digits.	>> 290/7 ans = 4.1429e+001
format long e	Scientific notation with 15 decimal digits.	>> 290/7 ans = 4.142857142857143e+001
format short g	Best of 5-digit fixed or floating point.	>> 290/7 ans = 41.429
format long g	Best of 15-digit fixed or floating point.	>> 290/7 ans = 41.4285714285714
format bank	Two decimal digits.	>> 290/7 ans = 41.43
format compact	Eliminates empty lines to allow more lines with information displayed on the screen.	
format loose	Adds empty lines (opposite of compact).	

1.5 ELEMENTARY MATH BUILT-IN FUNCTIONS

In addition to basic arithmetic operations, expressions in MATLAB can include functions. MATLAB has a very large library of built-in functions. A function has a name and an argument in parentheses. For example, the function that calculates the square root of a number is sqrt(x). Its name is sqrt, and the argument is x. When the function is used, the argument can be a number, a variable that has been assigned a numerical value (explained in Section 1.6), or a computable expression that can be made up of numbers and/or variables. Functions can also be included in arguments, as well as in expressions. Tutorial 1-2 shows examples of using the function `sqrt(x)` when MATLAB is used as a calculator with scalars.

Tutorial 1-2: Using the `sqrt` built-in function.

```
>> sqrt(64)
```
Argument is a number.
```
ans =
     8
>> sqrt(50+14*3)
```
Argument is an expression.
```
ans =
    9.5917
>> sqrt(54+9*sqrt(100))
```
Argument includes a function.
```
ans =
    12
>> (15+600/4)/sqrt(121)
```
Function is included in an expression.
```
ans =
    15
>>
```

Lists of some commonly used elementary MATLAB mathematical built-in functions are given in Tables 1-3 through 1-5. A complete list of functions organized by name of category can be found in the Help Window.

Table 1-3: Elementary math functions

Function	Description	Example

`sqrt(x)`	Square root.	`>> sqrt(81)`
		`ans =`
		`     9`

`nthroot(x,n)`	Real nth root or a real number x. (If x is negative n must be an odd integer.)	`>> nthroot(80,5)`
		`ans =`
		`    2.4022`

`exp(x)`	Exponential (e^x).	`>> exp(5)`
		`ans =`
		`    148.4132`

`abs(x)`	Absolute value.	`>> abs(-24)`
		`ans =`
		`    24`

`log(x)`	Natural logarithm. Base e logarithm (ln).	`>> log(1000)`
		`ans =`
		`    6.9078`

`log10(x)`	Base 10 logarithm.	`>> log10(1000)`
		`ans =`
		`    3.0000`

Table 1-3: Elementary math functions (Continued)

Function	Description	Example

`factorial(x)`	The factorial function $x!$ (x must be a positive integer.)	`>> factorial(5)`
		`ans =`
		`    120`

Table 1-4: Trigonometric math functions

Function	Description	Example

`sin(x)`
`sind(x)`
	Sine of angle x (x in radians).
	Sine of angle x (x in degrees).
		`>> sin(pi/6)`
		`ans =`
		`    0.5000`

`cos(x)`
`cosd(x)`
	Cosine of angle x (x in radians).
	Cosine of angle x (x in degrees).
		`>> cosd(30)`
		`ans =`
		`    0.8660`

`tan(x)`
`tand(x)`
	Tangent of angle x (x in radians).
	Tangent of angle x (x in degrees).
		`>> tan(pi/6)`
		`ans =`
		`    0.5774`

`cot(x)`
`cotd(x)`
	Cotangent of angle x (x in radians).
	Cotangent of angle x (x in radians).
		`>> cotd(30)`
		`ans =`
		`    1.7321`

The inverse trigonometric functions are `asin(x)`, `acos(x)`, `atan(x)`, `acot(x)` for the angle in radians, and `asind(x)`, `acosd(x)`, `atand(x)`, `acotd(x)` for the angle in degrees. The hyperbolic trigonometric functions are `sinh(x)`, `cosh(x)`, `tanh(x)`, and `coth(x)`. The previous table uses `pi` which is equal to π (see Section 1.6.3).

Table 1-5: Rounding functions

Function	Description	Example

`round(x)`	Round to the nearest integer.	`>> round(17/5)`
		`ans =`
		`    3`

`fix(x)`	Round towards zero.	`>> fix(13/5)`
		`ans =`
		`    2`

`ceil(x)`	Round towards infinity.	`>> ceil(11/5)`
		`ans =`
		`    3`

`floor(x)`	Round towards minus infinity.	`>> floor(-9/4)`
		`ans =`
		`    -3`

`rem(x,y)`	Returns the remainder after x is divided by y.	`>> rem(13,5)`
		`ans =`
		`    3`

Table 1-5: Rounding functions

Function	Description	Example
sign(x)	Signum function. Returns 1 if $x > 0$, -1 if $x < 0$, and 0 if $x = 0$.	>> sign(5) ans = 1

1.6 DEFINING SCALAR VARIABLES

A variable is a name made of a letter or a combination of several letters (and digits) that is assigned a numerical value. Once a variable is assigned a numerical value, it can be used in mathematical expressions, in functions, and in any MATLAB statements and commands. A variable is actually a name of a memory location. When a new variable is defined, MATLAB allocates an appropriate memory space where the variable's assignment is stored. When the variable is used the stored data is used. If the variable is assigned a new value the content of the memory location is replaced. (In Chapter 1 we only consider variables that are assigned numerical values that are scalars. Assigning and addressing variables that are arrays is discussed in Chapter 2.)

1.6.1 The Assignment Operator

In MATLAB the $=$ sign is called the assignment operator. The assignment operator assigns a value to a variable.

> Variable_name = A numerical value, or a computable expression

• The left-hand side of the assignment operator can include only one variable name. The right-hand side can be a number, or a computable expression that can include numbers and/or variables that were previously assigned numerical values. When the **Enter** key is pressed the numerical value of the right-hand side is assigned to the variable, and MATLAB displays the variable and its assigned value in the next two lines.

The following shows how the assignment operator works:

```
>> x=15

x =

    15

>> x=3*x-12

x =

    33

>>
```

The number 15 is assigned to the variable x.

MATLAB displays the variable and its assigned value.

A new value is assigned to x. The new value is 3 times the previous value of x minus 12.

The last statement ($x = 3x - 12$) illustrates the difference between the assignment operator and the equal sign. If in this statement the $=$ sign meant equal, the value of x would be 6 (solving the equation for x).

The use of previously defined variables to define a new variable is demonstrated next.

```
>> a=12
```
Assign 12 to a.
```
a =
    12
>> B=4
```
Assign 4 to B.
```
B =
     4
>> C=(a-B)+40-a/B*10
```
Assign the value of the expression on the right-hand side to the variable C.
```
C =
    18
```

- If a semicolon is typed at the end of the command then, when the **Enter** key is pressed, MATLAB does not display the variable with its assigned value (the variable still exists and is stored in memory).

- If a variable already exists, typing the variable's name and pressing the **Enter** key will display the variable and its value in the next two lines.

For example, the last demonstration is repeated below using semicolons:

```
>> a=12;
>> B=4;
>> C=(a-B)+40-a/B*10;
```
The variables a, B, and C are defined but are not displayed since a semicolon is typed at the end of each statement.
```
>> C
C =
    18
```
The value of the variable C is displayed by typing the name of the variable.

- Several assignments can be typed in the same line. The assignments must be separated with a comma (spaces can be added after the comma). When the **Enter** key is pressed, the assignments are executed from left to right and the variables and their assignments are displayed. A variable is not displayed if a semicolon is typed instead of a comma. For example, the assignments of the variables a, B, and C above can all be done in the same line.

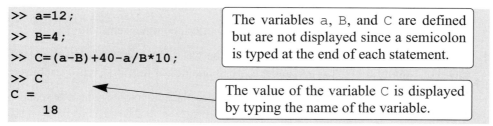

```
>> a=12, B=4; C=(a-B)+40-a/B*10
a =
    12
C =
    18
```
The variable B is not displayed because a semicolon is typed at the end of the assignment.

- A variable that already exists can be reassigned a new value. For example:

```
>> ABB=72;          A value of 72 is assigned to the variable ABB.
>> ABB=9;           A new value of 9 is assigned to the variable ABB.
>> ABB
ABB =               The current value of the variable is dis-
      9             played when the name of the variable is
>>                  typed and the Enter key is pressed.
```

- Once a variable is defined it can be used as an argument in functions. For example:

```
>> x=0.75;
>> E=sin(x)^2+cos(x)^2
E =
      1
>>
```

1.6.2 Rules About Variable Names

A variable can be named according to the following rules:
- Must begin with a letter.
- Can be up to 63 (in MATLAB 7) characters long (31 characters in MATLAB 6.0).
- Can contain letters, digits, and the underscore character.
- Cannot contain punctuation characters (e.g. period, comma, semicolon).
- MATLAB is case sensitive; it distinguishes between uppercase and lowercase letters. For example, AA, Aa, aA, and aa are the names of four different variables.
- No spaces are allowed between characters (use the underscore where a space is desired).
- Avoid using the names of a built-in function for a variable (i.e. avoid using: cos, sin, exp, sqrt, etc.). Once a function name is used to define a variable, the function cannot be used.

1.6.3 Predefined Variables and keywords

There are seventeen words, called keywords, that are reserved by MATLAB for various purposes, and cannot be used as variable names. These words are:

```
break     case     catch    continue   else     elseif    end
for       function global    if        otherwise persistent
return    switch    try      while
```

When typed, these words appear in blue. An error message is displayed if the user tries to use a keyword as a variable name. (The keywords can be displayed by typing the command `iskeyword`)

A number of frequently used variables are already defined when MATLAB is started. Some of the predefined variables are:

ans   A variable that has the value of the last expression that was not assigned to a specific variable (see Tutorial 1-1). If the user does not assign the value of an expression to a variable, MATLAB automatically stores the result in `ans`.

pi   The number π.

eps   The smallest difference between two numbers. Equal to $2^{\wedge}(-52)$, which is approximately 2.2204e–016.

inf   Used for infinity.

i   Defined as $\sqrt{-1}$, which is: 0 + 1.0000i.

j   Same as i.

NaN   Stands for Not-a-Number. Used when MATLAB cannot determine a valid numeric value. For example 0/0.

The predefined variables can be redefined to have any other value. The variables `pi`, `eps`, and `inf`, are usually not redefined since they are frequently used in many applications. Other predefined variables like i and j are sometime redefined (commonly in association with loops) when complex numbers are not involved in the application.

1.7 USEFUL COMMANDS FOR MANAGING VARIABLES

The following are commands that can be used to eliminate variables or to obtain information about variables that have been created. When these commands are typed in the Command Window and the **Enter** key is pressed, they either provide information, or they perform a task as listed below.

Command	Outcome
clear	Removes all variables from the memory.
clear x y z	Removes only variables x, y, and z from the memory.
who	Displays a list of the variables currently in the memory.
whos	Displays a list of the variables currently in the memory and their size together with information about their bytes and class (see Section 4.1).

1.8 SCRIPT FILES

So far all the commands were typed in the Command Window and were executed when the **Enter** key was pressed. Although every MATLAB command can be executed in this way, using the Command Window to execute a series of commands—especially if they are related to each other (a program)—is not convenient and may be difficult or even impossible. The commands in the Command Window cannot be saved and executed again. In addition, the Command Window is not interactive. This means that every time the **Enter** key is pressed only the last command is executed, and everything executed before is unchanged. If a change or a correction is needed in a command that was previously executed and the results of this command are used in commands that follow, all the commands have to be entered and executed again.

A different (better) way of executing commands with MATLAB is first to create a file with a list of commands (program), save it, and then run (execute) the file. When the file runs, the commands it contains are executed in the order that they are listed. If needed, the commands in the file can be corrected or changed and the file can be saved and run again. Files that are used for this purpose are called script files.

IMPORTANT NOTE: This section covers only the minimum that is required in order to run simple programs. This will allow the student to use script files when practicing the material that is presented in this and the next two chapters (instead of typing repeatedly in the Command Window). Script files are considered again in Chapter 4 where many additional topics that are essential for understanding MATLAB and writing programs in script file are covered.

1.8.1 Notes About Script Files

- A script file is a sequence of MATLAB commands, also called a program.

- When a script file runs (is executed), MATLAB executes the commands in the order they are written just as if they were typed in the Command Window.

- When a script file has a command that generates an output (e.g. assignment of a value to a variable without semicolon at the end), the output is displayed in the Command Window.

- Using a script file is convenient because it can be edited (corrected and/or changed) and executed many times.

- Script files can be typed and edited in any text editor and then pasted into the MATLAB editor.

- Script files are also called M-files because the extension .m is used when they are saved.

1.8.2 Creating and Saving a Script File

In MATLAB script files are created and edited in the Editor/Debugger Window. This window is opened from the Command Window. In the **File** menu, select **New**, and then select **M-file**. An open Editor/Debugger Window is shown in Figure 1-6.

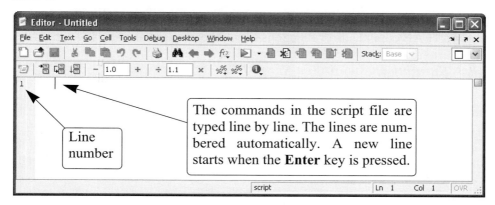

Figure 1-6: The Editor/Debugger Window.

Once the window is open, the commands of the script file are typed line by line. MATLAB automatically numbers a new line every time the **Enter** key is pressed. The commands can also be typed in any text editor or word processor program and then copied and pasted in the Editor/Debugger Window. An example of a short program typed in the Editor/Debugger Window is shown in Figure 1-7. The first few lines in a script file are typically comments (which are not executed since the first character in the line is %) that describe the program written in the script file.

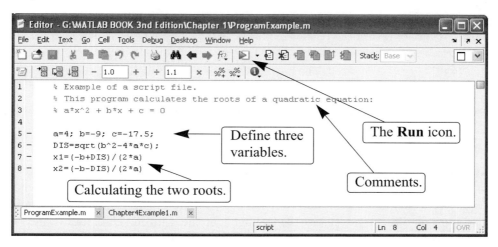

Figure 1-7: A program typed in the Editor/Debugger Window.

Before a script file can be executed it has to be saved. This is done by choosing **Save As**... from the **File** menu, selecting a location (many students save to a flush drive which appears in the directory as `Drive(F:)` or `(G:)`), and entering a name for the file. When saved, MATLAB adds the extension .m to the mane. The rules for naming a script file follow the rules of naming a variable (must begin with a letter, can include digits and underscore, <u>no spaces</u>, and be up to 63 characters long). The names of user-defined variables, predefined variables, and MATLAB commands or functions should not be used as names of script files.

1.8.3 Running (Executing) a Script File

A script file can be executed either directly from the Editor Window by clicking on the **Run** icon (see Figure 1-7), or by typing the file name in the Command Window and then pressing the **Enter** key. To be executed, MATLAB needs to know where the file is saved. The file will be executed if the directory where the file is saved is the current directory of MATLAB or if the directory is listed in the search path, as explained next.

1.8.4 Current Directory

The current directory is shown in the "Current Directory" field in the desktop toolbar of the Command Window, as shown in Figure 1-8. If an attempt is made to execute a script file by clicking on the **Run** icon (in the Editor Window) while the current directory is not the directory where the script file is saved, then the prompt shown in Figure 1-9 will open. The user can then change the current directory to

Figure 1-8: The Current Directory field in the Command Window.

the directory where the script file is saved, or add it to the search path. Once two or more different current directories are used in a session, it is possible to switch from one to another in the **Current Directory** field in the Command Window. The current directory can also be changed in the Current Directory Window, shown in Figure 1-10, that can be opened from the **Desktop** menu. The Current Directory can be changed by choosing the drive and folder where the file is saved.

Figure 1-9: Changing the Current Directory.

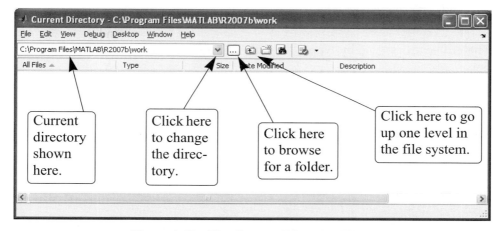

Figure 1-10: The Current Directory Window.

An alternative simple way to change the current directory is to use the cd command in the Command Window. To change the current directory to a different drive, type cd, space, and then the name of the directory followed by a colon : and press the **Enter** key. For example, to change the current directory to drive F (e.g., the flush drive) type cd F:. If the script file is saved in a folder within a drive, the path to that folder has to be specified. This is done by typing the path as a string in the cd command. For example, cd('F:\Chapter 1') sets the path to the folder Chapter 1 in drive F. The following example shows how the current directory is changed to be drive F. Then, the script file from Figure 1-7, which was saved in drive F as: ProgramExample.m, is executed by typing the name of the file and pressing the **Enter** key.

1.9 EXAMPLES OF MATLAB APPLICATIONS

Sample Problem 1-1: Trigonometric identity

A trigonometric identity is given by:

$$\cos^2\frac{x}{2} = \frac{\tan x + \sin x}{2\tan x}$$

Verify that the identity is correct by calculating each side of the equation, substituting $x = \frac{\pi}{5}$.

Solution

The problem is solved by typing the following commands in the Command Window.

```
>> x=pi/5;                                    Define x.

>> LHS=cos(x/2)^2                             Calculate the left-hand side.

LHS =
    0.9045

>> RHS=(tan(x)+sin(x))/(2*tan(x))            Calculate the right-hand side.

RHS =
    0.9045
```

Sample Problem 1-2: Geometry and trigonometry

Four circles are placed, as shown in the figure. At each point that two circles are in contact they are tangent to each other. Determine the distance between the centers C_2 and C_4.

The radii of the circles are:
$R_1 = 16\,$mm, $R_2 = 6.5\,$mm, $R_3 = 12\,$mm, and $R_4 = 9.5\,$mm.

Solution

The lines that connect the centers of the circles create four triangles. In two of the triangles, $\Delta C_1C_2C_3$ and $\Delta C_1C_3C_4$, the lengths of all the sides are known. This information is used to calculate the angles γ_1 and γ_2 in these triangles by using the law of cosines. For example, γ_1 is calculated from:

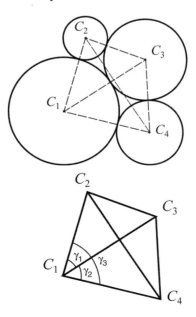

$$(C_2C_3)^2 = (C_1C_2)^2 + (C_1C_3)^2 - 2(C_1C_2)(C_1C_3)\cos\gamma_1$$

Next, the length of the side C_2C_4 is calculated by considering the triangle $\Delta C_1C_2C_4$. This is done, again, by using the law of cosines (the lengths C_1C_2 and C_1C_4 are known and the angle γ_3 is the sum of the angles γ_1 and γ_2).

The problem is solved by writing the following program in a script file:

```
% Solution of Sample Problem 1-2
R1=16; R2=6.5; R3=12; R4=9.5;
R1=16; R2=6.5; R3=12; R4=9.5;                 Define the R's.
C1C2=R1+R2; C1C3=R1+R3; C1C4=R1+R4;           Calculate the
                                              lengths of the sides.
C2C3=R2+R3; C3C4=R3+R4;
Gama1=acos((C1C2^2+C1C3^2-C2C3^2)/(2*C1C2*C1C3));
Gama2=acos((C1C3^2+C1C4^2-C3C4^2)/(2*C1C3*C1C4));
Gama3=Gama1+Gama2;                            Calculate γ₁, γ₂, and γ₃.

C2C4=sqrt(C1C2^2+C1C4^2-2*C1C2*C1C4*cos(Gama3))
                                              Calculate the length of
                                              side C₂C₄.
```

When the script file is executed, the following (the value of the variable C2C4) is displayed in the Command Window:

```
C2C4 =
   33.5051
```

Sample Problem 1-3: Heat transfer

An object with an initial temperature of T_0 that is placed at time $t = 0$ inside a chamber that has a constant temperature of T_s, will experience a temperature change according to the equation:

$$T = T_s + (T_0 - T_s)e^{-kt}$$

where T is the temperature of the object at time t, and k is a constant. A soda can at a temperature of 120°F (was left in the car) is placed inside a refrigerator where the temperature is 38°F. Determine, to the nearest degree, the temperature of the can after three hours. Assume $k = 0.45$. First define all the variables and then calculate the temperature using one MATLAB command.

Solution

The problem is solved by typing the following commands in the Command Window.

```
>> Ts=38;   T0=120; k=0.45; t=3;

>> T=round(Ts+(T0-Ts)*exp(-k*t))

T =
    59
```

Round to the nearest integer.

Sample Problem 1-4:   Compounded interest

The balance B of a savings account after t years when a principal P is invested at an annual interest rate r and the interest is compounded n times a year is given by:

$$B = P\left(1 + \frac{r}{n}\right)^{nt} \quad (1)$$

If the interest is compounded yearly, the balance is given by:

$$B = P(1 + r)^t \quad (2)$$

In one account $5,000 is invested for 17 years in an account where the interest is compounded yearly. In a second account $5,000 is invested in an account in which the interest is compounded monthly. In both accounts the interest rate is 8.5%. Use MATLAB to determine how long (in years and months) it would take for the balance in the second account to be the same as the balance of the first account after 17 years.

Solution

Follow these steps:
(*a*) Calculate B for $5,000 invested in a yearly compounded interest account after 17 years using Equation (2).
(*b*) Calculate the t for the B calculated in part (*a*), from the monthly compounded interest formula, Equation (1).
(*c*) Determine the number of years and months that correspond to t.
    The problem is solved by writing the following program in a script file:

```
% Solution of Sample Problem 1-4
P=5000;   r=0.085;   ta=17; n=12;
B=P*(1+r)^ta                           Step (a): Calculate B from Eq. (2).
t=log(B/P)/(n*log(1+r/n))              Step (b): Solve Eq. (1)
                                       for t, and calculate t.

years=fix(t)                           Step (c): Determine the number of years.
months=ceil((t-years)*12)             Determine the number of months.
```

When the script file is executed, the following (the value of the variable C2C4) is displayed in the Command Window:

```
>> format short g
B =
          20011

t =
          16.374

years =
       16

months =
        5
```

> The values of the variables B, t, years, and months are displayed (since a semicolon was not typed at the end of the commands that calculates the values).

1.10 PROBLEMS

The following problems can be solved by writing commands in the Command Window, or by writing a program in a script file and then executing the file.

1. Calculate:

   a) $\dfrac{28.5 \cdot 3^3 - \sqrt{1500}}{11^2 + 37.3}$

   b) $\left(\dfrac{7}{3}\right)^2 \cdot 4^3 \cdot 18 - \dfrac{6^7}{(9^3 - 652)}$

2. Calculate:

   a) $23\left(-8 + \dfrac{\sqrt{607}}{3}\right) + \left(\dfrac{40}{8} + 4.7^2\right)^2$

   b) $509^{1/3} - 4.5^2 + \dfrac{\ln 200}{1.5} + 75^{1/2}$

3. Calculate:

   a) $\dfrac{(24 + 4.5^3)}{e^{4.4} - \log_{10}(12560)}$

   b) $\dfrac{2}{0.036} \cdot \dfrac{(\sqrt{250} - 10.5)^2}{e^{-0.2}}$

4. Calculate:

a) $\cos\left(\dfrac{5\pi}{6}\right)\sin^2\left(\dfrac{7\pi}{8}\right) + \dfrac{\tan\left(\dfrac{\pi}{6}\ln 8\right)}{\sqrt{7}+2}$

b) $\cos^2\left(\dfrac{3\pi}{5}\right) + \dfrac{\tan\left(\dfrac{\pi\ln 6}{5}\right)}{8\cdot\dfrac{7}{2}}$

5. Define the variable x as $x = 9.75$, then evaluate:

a) $4x^3 - 14x^2 - 6.32x + 7.3$

b) $\dfrac{e^{\sqrt{3}}}{\sqrt[3]{0.02\cdot 3.1^2}}$

c) $\log_{10}(x^2 - x^3)^2$

6. Define the variables x and z as $x = 5.3$, and $z = 7.8$, then evaluate:

a) $\dfrac{xz}{(x/z)^2} + 14x^2 - 0.8z^2$

b) $x^2 z - z^2 x + \left(\dfrac{x}{z}\right)^2 - \left(\dfrac{z}{x}\right)^{1/2}$

7. Define the variables a, b, c, and d as:
$a = -18.2$, $b = 6.42$, $c = a/b$ and $d = 0.5(cb + 2a)$, and then evaluate:

a) $d - \dfrac{a+b}{c} + \dfrac{(a+d)^2}{\sqrt{|abc|}}$

b) $\ln[(c-d)(b-a)] + \dfrac{(a+b+c+d)}{(a-b-c-d)}$

8. A sphere has a radius of 15 cm.
   a) Determine the length of a side of a cube that has the same surface area as the sphere.
   b) Determine the length of a side of a cube that has the same volume as the sphere.

9. Calculate the volume of a sphere that has a surface area of 200 in^2 in two ways:
   a) First calculate the radius of the sphere r, then substitute the radius in the formula of the volume.
   b) Write one command.

10. Two trigonometric identities are given by:

    a) $\sin 3x = 3\sin x - 4\sin^3 x$

    b) $\sin\dfrac{x}{2} = \sqrt{\dfrac{1 - \cos x}{2}}$

    For each part, verify that the identity is correct by calculating each side of the equation, substituting $x = \dfrac{7}{20}\pi$.

11. Two trigonometric identities are given by:

    a) $\tan 3x = \dfrac{3\tan x - \tan^3 x}{1 - 3\tan^2 x}$

    b) $\tan\dfrac{x}{2} = \dfrac{\sin x}{1 + \cos x}$

    For each part, verify that the identity is correct by calculating the values of the left and right sides of the equation, substituting $x = 27°$.

12. Define two variables: $alpha = 5\pi/9$, $beta = \pi/7$. Using these variables, show that the following trigonometric identity is correct by calculating the value of the left- and right-hand sides of the equation.

$$\sin\alpha\sin\beta = \frac{1}{2}[\cos(\alpha - \beta) - \cos(\alpha + \beta)]$$

13. Given: $\displaystyle\int \sin^2 x\,dx = \frac{1}{2}x - \frac{1}{4}\sin 2x$. Use MATLAB to calculate the following definite integral: $\displaystyle\int_{\frac{\pi}{3}}^{\frac{3\pi}{4}} \sin^2 x\,dx$.

14. In the triangle shown $a = 21\,\text{cm}$, $b = 45\,\text{cm}$, and $c = 60\,\text{cm}$. Define a, b, and c as variables, and then:

    a)   Calculate the angle γ (in degrees) by substituting the variables in the Law of Cosines.

    (The Law of Cosines: $c^2 = a^2 + b^2 - 2ab\cos\gamma$)

    b)   Calculate the angles α and β (in degrees) using the Law of Sines.

    c)   Check that the sum of the angles is $180°$.

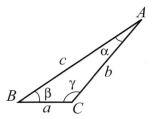

15. In the right triangle shown $a = 15$ cm, and $b = 35$ cm. Define a and c as variables, and then:

    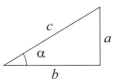

    a) Using the Pythagorean Theorem, calculate c by typing one line in the Command Window.

    b) Using c from part a), and the acosd function, calculate the angle α in degrees, typing one line in the Command Window.

16. The distance d from a point (x_0, y_0) to a line $Ax + By + C = 0$ is given by:

$$d = \frac{|Ax_0 + By_0 + C|}{\sqrt{A^2 + B^2}}$$

    Determine the distance of the point $(3, -4)$ from the line $2x - 7y - 10 = 0$. First define the variables A, B, C, x_0, and y_0, and then calculate d. (Use the abs and sqrt functions.)

17. Eggs are packed in containers such that 18 are placed in each container. Determine how many containers are needed to pack 634 eggs. Use MATLAB built-in function ceil.

18. Define the following two variables:

    CD_price = $13.95   Book_price = $44.95

    Then change the display format to bank and calculate the following by typing one command:

    a) The cost of three CDs and five books.

    b) The same as part a), but add 5.75% sale tax.

    c) The same as part b) but round the total cost to the nearest dollar.

19. The number of combinations $C_{n,r}$ of taking r objects out of n objects is given by:

$$C_{n,r} = \frac{n!}{r!(n-r)!}$$

    A basketball team has 12 players on the roster. Determine how many different teams of 5 players can be selected out of the 12 players. (Use the built-in function factorial.)

20. The formula for changing the base of a logarithm is:

$$\log_a N = \frac{\log_b N}{\log_b a}$$

    a) Use MATLAB's function log(x) to calculate $\log_5 281$.

    b) Use MATLAB's function log10(x) to calculate $\log_7 1054$.

21. Radioactive decay is modeled with the exponential function $f(t) = f(0)e^{kt}$, where t is time, $f(0)$ is the amount of material at $t = 0$, $f(t)$ is the amount of material at time t, and k is a constant. Gallium-67, which has a half-life of 3.261 days, is used for tracing cancer. If 100 milligrams are present at $t = 0$, determine the amount (rounded to the nearest tenth of a milligram) that is left after 7 days. Solve the problem by writing a program in a script file. The program first determines the constant k, then calculates $f(7)$, and finally rounds the answer to the nearest tenth of a milligram.

22. Fractions can be added by using the smallest common denominator. For example, the smallest common denominator of 1/4 and 1/10 is 20. Use the MATLAB Help Window to find a MATLAB built-in function that determines the least common multiple of two numbers. Then use the function to show that the least common multiple of:
    a)   4 and 14 is 28.
    b)   8 and 42 is 168.

23. The magnitude M of an earthquake on the Richter scale is given by $M = \frac{2}{3}\log_{10}\left(\frac{E}{E_0}\right)$, where E is the energy released by the earthquake, and $E_0 = 10^{4.4}$ Joules is a constant (energy of a small reference earthquake). Determine how many times more energy is released from an earthquake that registers 7.1 on the Richter scale than an earthquake that registers 6.9.

24. The balance B of a savings account after t years when a principal P is invested at an annual interest rate r and the interest is compounded yearly is given by $B = P(1 + r)^t$. If the interest is compounded continuously the balance is given by $B = Pe^{rt}$. In one account \$20,000 is invested for 18 years in an account where the interest is compounded yearly. In a second account \$5,000 is invested in an account in which the interest is compounded continuously. In both accounts the interest rate is 8.5%. Use MATLAB to determine how long in years and days (e.g. 17 years and 251 days) it would take for the balance in the second account to be the same as the balance of the first account after 18 years.

25. The temperature dependence of vapor pressure p can be estimated by the Antoine equation:
$$\ln(p) = A - \frac{B}{C + T}$$
where ln is the natural logarithm, p is in mm Hg, T is in Kelvin, and A, B, and C are material constants. For toluene ($C_6H_5CH_3$) in the temperature range from 280 to 410 K the material constants are: $A = 16.0137$, $B = 3096.52$, and $C = -53.67$. Calculate the vapor pressure of toluene at 315 and 405 K.

26. Sound level L_P in units of decibels (dB) is determined by:

$$L_P = 20\log_{10}\left(\frac{p}{p_0}\right)$$

where p is the sound pressure of the sound, and $p_0 = 20 \times 10^{-6}$ Pa is a reference sound pressure (the sound pressure when $L_P = 0$ dB). Determine the sound pressure of 90 dB noise (noise generated by a passing truck). By how many times the sound pressure of the truck is larger (louder) than the sound pressure during a normal conversation where the loudness is 65 dB.

Chapter 2
Creating Arrays

The array is a fundamental form that MATLAB uses to store and manipulate data. An array is a list of numbers arranged in rows and/or columns. The simplest array (one-dimensional) is a row, or a column of numbers. A more complex array (two-dimensional) is a collection of numbers arranged in rows and columns. One use of arrays is to store information and data, as in a table. In science and engineering, one-dimensional arrays frequently represent vectors, and two-dimensional arrays often represent matrices. Chapter 2 shows how to create and address arrays while Chapter 3 shows how to use arrays in mathematical operations. In addition to arrays that are made of numbers, arrays in MATLAB can also be made of a list of characters, which are called strings. Strings are discussed in Section 2.10.

2.1 CREATING A ONE-DIMENSIONAL ARRAY (VECTOR)

A one-dimensional array is a list of numbers that is placed in a row or a column. One example is the representation of the position of a point in space in a three-dimensional Cartesian coordinate system. As shown in Figure 2-1, the position of point A is defined by a list of three numbers 2, 4, and 5, which are the coordinates of the point.

The position of point A can be expressed in terms of a position vector:

$$\mathbf{r}_A = 2\mathbf{i} + 4\mathbf{j} + 5\mathbf{k}$$

where $\mathbf{i}$, $\mathbf{j}$, and $\mathbf{k}$ are unit vectors in the direction of the x, y, and z axis, respectively. The numbers 2, 4, and 5 can be used to define a row or a column vector.

Any list of numbers can be set up as a vector. For example, Table 2-1 contains population growth data that can be used to create two lists of numbers; one of the years and the other of the popula-

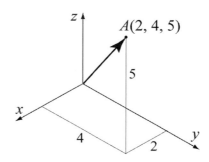

Figure 2-1: Position of a point.

tion. Each list can be entered as elements in a vector with the numbers placed in a row or in a column.

Table 2-1: Population data

Year	1984	1986	1988	1990	1992	1994	1996
Population (Millions)	127	130	136	145	158	178	211

In MATLAB, a vector is created by assigning the elements of the vector to a variable. This can be done in several ways depending on the source of the information that is used for the elements of the vector. When a vector contains specific numbers that are known (like the coordinates of point *A*), the value of each element is entered directly. Each element can also be a mathematical expression that can include predefined variables, numbers, and functions. Often, the elements of a row vector are a series of numbers with constant spacing. In such cases the vector can be created with MATLAB commands. A vector can also be created as the result of mathematical operations as explained in Chapter 3.

Creating a vector from a known list of numbers:

The vector is created by typing the elements (numbers) inside square brackets [ ].

```
variable_name = [ type vector elements ]
```

Row vector: To create a row vector type the elements with a space or a comma between the elements inside the square brackets.

Column vector: To create a column vector type the left square bracket [and then enter the elements with a semicolon between them, or press the **Enter** key after each element. Type the right square bracket] after the last element.

Tutorial 2-1 shows how the data from Table 2-1 and the coordinates of point *A* are used to create row and column vectors.

Tutorial 2-1: Creating vectors from given data.

```
>> yr=[1984 1986 1988 1990 1992 1994 1996]
```
   The list of years is assigned to a row vector named yr.
```
yr =
      1984      1986      1988      1990      1992      1994      1996
>> pop=[127;  130;  136;  145;  158;  178;  211]
```
   The population data is assigned to a column vector named pop.
```
pop =
      127
      130
      136
      145
      158
```

Tutorial 2-1: Creating vectors from given data. (Continued)

```
      178
      211
>> pntAH=[2,   4,   5]

pntAH =
      2       4      5

>> pntAV=[2
4
5]

pntAV =
      2
      4
      5
>>
```

> The coordinates of point *A* are assigned to a row vector called `pntAH`.

> The coordinates of point *A* are assigned to a column vector called `pntAV`. (The **Enter** key is pressed after each element is typed.)

Creating a vector with constant spacing by specifying the first term, the spacing, and the last term:

In a vector with constant spacing the difference between the elements is the same. For example, in the vector: $v = 2\ 4\ 6\ 8\ 10$, the spacing between the elements is 2. A vector in which the first term is m, the spacing is q, and the last term is n is created by typing:

$$\text{variable\_name} = [\text{m:q:n}] \quad \text{or} \quad \text{variable\_name} = \text{m:q:n}$$

(The brackets are optional.)

Some examples are:

```
>> x=[1:2:13]

x =
      1      3      5      7      9     11     13
>> y=[1.5:0.1:2.1]

y =
   1.5000   1.6000   1.7000   1.8000   1.9000   2.0000   2.1000


>> z=[-3:7]

z =
     -3     -2     -1      0      1      2      3      4      5      6
7
>> xa=[21:-3:6]
```

> First element 1, spacing 2, last element 13.

> First element 1.5, spacing 0.1, last element 2.1.

> First element –3, last term 7. If spacing is omitted, the default is 1.

> First element 21, spacing –3, last term 6.

```
xa =
      21    18    15    12     9     6
>>
```

- If the numbers m, q, and n are such that the value of n can not be obtained by adding q's to m, then (for positive n) the last element in the vector will be the last number that does not exceed n.

Creating a vector with constant spacing by specifying the first and last terms, and the number of terms:

A vector in which the first element is *xi*, the last element is *xf*, and the number of elements is *n* is created by typing the `linspace` command (MATLAB determines the correct spacing):

$$\boxed{\texttt{variable\_name = linspace(xi,xf,n)}}$$

Some examples are:

```
>> va=linspace(0,8,6)        6 elements, first element 0, last element 8.
va =
        0    1.6000    3.2000    4.8000    6.4000    8.0000
>> vb=linspace(30,10,11)     11 elements, first element 30, last element 10.
vb =
      30    28    26    24    22    20    18    16    14    12
10
>> u=linspace(49.5,0.5)      First element 49.5, last element 0.5.

                             When the number of elements is
u =                          omitted, the default is 100.
  Columns 1 through 10
   49.5000   49.0051   48.5101   48.0152   47.5202   47.0253
  46.5303   46.0354   45.5404   45.0455
. . . . . . . . . .          100 elements are displayed.
  Columns 91 through 100
    4.9545    4.4596    3.9646    3.4697    2.9747    2.4798
  1.9848    1.4899    0.9949    0.5000
>>
```

2.2 *CREATING A TWO-DIMENSIONAL ARRAY (MATRIX)*

A two-dimensional array, also called a matrix, has numbers in rows and columns. Matrices can be used to store information like in a table. Matrices play an important role in linear algebra and are used in science and engineering to describe many physical quantities.

In a square matrix the number of rows and columns is equal. For example, the matrix:

$$\begin{matrix} 7 & 4 & 9 \\ 3 & 8 & 1 \\ 6 & 5 & 3 \end{matrix} \quad 3 \times 3 \text{ matrix}$$

is square, with three rows and three columns. In general, the number of rows and columns can be different. For example, the matrix:

$$\begin{matrix} 31 & 26 & 14 & 18 & 5 & 30 \\ 3 & 51 & 20 & 11 & 43 & 65 \\ 28 & 6 & 15 & 61 & 34 & 22 \\ 14 & 58 & 6 & 36 & 93 & 7 \end{matrix} \quad 4 \times 6 \text{ matrix}$$

has four rows and six columns. A $m \times n$ matrix has m rows and n columns, and m by n is called the size of the matrix.

A matrix is created by assigning the elements of the matrix to a variable. This is done by typing the elements, row by row, inside square brackets [ ]. First type the left bracket [, then type the first row separating the elements with spaces or commas. To type the next row type a semicolon or press **Enter**. Type the right bracket] at the end of the last row.

> variable_name=[1st row elements; 2nd row elements; 3rd row elements; ..... ; last row elements]

The elements that are entered can be numbers or mathematical expressions that may include numbers, predefined variables, and functions. All the rows must have the same number of elements. If an element is zero, it has to be entered as such. MATLAB displays an error message if an attempt is made to define an incomplete matrix. Examples of matrices defined in different ways are shown in Tutorial 2-2.

Tutorial 2-2: Creating matrices.

```
>> a=[5   35   43;   4   76   81;   21   32   40]
a =
        5        35        43
        4        76        81
       21        32        40
>> b = [7   2   76   33   8
1   98   6   25   6
5   54   68   9   0]
b =
        7         2        76        33         8
        1        98         6        25         6
        5        54        68         9         0
>> cd=6;  e=3;  h=4;
```

A semicolon is typed before a new line is entered.

The **Enter** key is pressed before a new line is entered.

Three variables are defined.

Tutorial 2-2: Creating matrices. (Continued)

```
>> Mat=[e, cd*h, cos(pi/3); h^2, sqrt(h*h/cd), 14]
Mat =
      3.0000     24.0000      0.5000
     16.0000      1.6330     14.0000
>>
```

Elements are defined by mathematical expressions.

Rows of a matrix can also be entered as vectors using the notation for creating vectors with constant spacing, or the `linspace` command. For example:

```
>> A=[1:2:11; 0:5:25; linspace(10,60,6); 67 2 43 68 4 13]
A =
       1       3       5       7       9      11
       0       5      10      15      20      25
      10      20      30      40      50      60
      67       2      43      68       4      13
>>
```

In this example the first two rows were entered as vectors using the notation of constant spacing, the third row was entered using the `linspace` command, and in the last row the elements were entered individually.

2.2.1 The `zeros`, `ones` and `eye` Commands

The `zeros(m,n)`, the `ones(m,n)`, and `eye(n)` commands can be used to create matrices that have elements with special values. The `zeros(m,n)` and the `ones(m,n)` commands create a matrix with m rows and n columns, in which all the elements are the numbers 0 and 1, respectively. The `eye(n)` command creates a square matrix with n rows and n columns in which the diagonal elements are equal to 1, and the rest of the elements are 0. This matrix is called the identity matrix. Examples are:

```
>> zr=zeros(3,4)
zr =
       0       0       0       0
       0       0       0       0
       0       0       0       0
>> ne=ones(4,3)
ne =
       1       1       1
       1       1       1
       1       1       1
       1       1       1
>> idn=eye(5)
```

```
idn =
     1     0     0     0     0
     0     1     0     0     0
     0     0     1     0     0
     0     0     0     1     0
     0     0     0     0     1
>>
```

Matrices can also be created as a result of mathematical operations with vectors and matrices. This topic is covered in Chapter 3.

2.3 NOTES ABOUT VARIABLES IN MATLAB

• All variables in MATLAB are arrays. A scalar is an array with one element, a vector is an array with one row, or one column, of elements, and a matrix is an array with elements in rows and columns.

• The variable (scalar, vector, or matrix) is defined by the input when the variable is assigned. There is no need to define the size of the array (single element for a scalar, a row or a column of elements for a vector, or a two-dimensional array of elements for a matrix) before the elements are assigned.

• Once a variable exists, as a scalar, vector, or a matrix, it can be changed to be any other size, or type, of variable. For example, a scalar can be changed to be a vector or a matrix, a vector can be changed to be a scalar, a vector of different length, or a matrix, and a matrix can be changed to have a different size, or to be reduced to a vector or a scalar. These changes are made by adding or deleting elements. This subject is covered in Sections 2.7 and 2.8.

2.4 THE TRANSPOSE OPERATOR

The transpose operator, when applied to a vector, switches a row (column) vector to a column (row) vector. When applied to a matrix, it switches the rows (columns) to columns (rows). The transpose operator is applied by typing a single quote ′ following the variable to be transposed. Examples are:

```
>> aa=[3  8  1]
aa =
     3     8     1
>> bb=aa'
bb =
     3
     8
     1
>> C=[2 55 14 8; 21 5 32 11; 41 64 9 1]
```

Define a row vector aa.

Define a column vector bb as the transpose of vector aa.

Define a matrix C with 3 rows and 4 columns.

```
C =
       2      55      14       8
      21       5      32      11
      41      64       9       1
>> D=C'

D =
       2      21      41
      55       5      64
      14      32       9
       8      11       1
>>
```

> Define a matrix D as the transpose of matrix C. (D has 4 rows and 3 columns.)

2.5 ARRAY ADDRESSING

Elements in an array (either vector or matrix) can be addressed individually or in subgroups. This is useful when there is a need to redefine only some of the elements, to use specific elements in calculations, or when a subgroup of the elements is used to define a new variable.

2.5.1 Vector

The address of an element in a vector is its position in the row (or column). For a vector named ve, ve(k) refers to the element in position k. The first position is 1. For example, if the vector ve has nine elements:

ve = 35 46 78 23 5 14 81 3 55

then

$ve(4) = 23$, $ve(7) = 81$, and $ve(1) = 35$.

A single vector element, $v(k)$, can be used just as a variable. For example, it is possible to change the value of only one element of a vector by reassigning a new value to a specific address. This is done by typing: $v(k) = value$. A single element can also be used as a variable in a mathematical expression. Examples are:

```
>> VCT=[35 46 78 23 5 14 81 3 55]          ◄——————  Define a vector.
VCT =
      35      46      78      23       5      14      81       3      55
>> VCT(4)                                   ◄——  Display the fourth element.
ans =
      23
                                            Assign a new value to
                                            the sixth element.
>> VCT(6)=273
VCT =                                       The whole vector is displayed.
      35      46      78      23       5     273      81       3      55
>> VCT(2)+VCT(8)
```

```
>> VCT(2)+VCT(8)
ans =
    49
>> VCT(5)^VCT(8)+sqrt(VCT(7))
ans =
   134
>>
```

Use the vector elements in mathematical expressions.

2.5.2 Matrix

The address of an element in a matrix is its position, defined by the row number and the column number where it is located. For a matrix assigned to a variable *ma*, *ma*(*k,p*) refers to the element in row *k* and column *p*.

For example, if the matrix is: $ma = \begin{bmatrix} 3 & 11 & 6 & 5 \\ 4 & 7 & 10 & 2 \\ 13 & 9 & 0 & 8 \end{bmatrix}$

then, *ma*(1,1) = 3, and *ma*(2,3) = 10.

As with vectors, it is possible to change the value of just one element of a matrix by assigning a new value to that element. Also, single elements can be used like variables in mathematical expressions and functions. Some examples are:

```
>> MAT=[3 11 6 5; 4 7 10 2; 13 9 0 8]     Create a 3 × 4 matrix.
MAT =
     3    11     6     5
     4     7    10     2
    13     9     0     8
>> MAT(3,1)=20                  Assign a new value to the (3,1) element.
MAT =
     3    11     6     5
     4     7    10     2
    20     9     0     8
>> MAT(2,4)-MAT(1,2)           Use elements in a mathematical expression.
ans =
    -9
```

2.6 USING A COLON : IN ADDRESSING ARRAYS

A colon can be used to address a range of elements in a vector or a matrix.

For a vector:

va(:)   Refers to all the elements of the vector *va* (either a row or a column vector).

va(*m:n*)   Refers to elements *m* through *n* of the vector *va*.

Example:

```
>> v=[4 15 8 12 34 2 50 23 11]                    A vector v is created.
v =
        4      15       8      12      34       2      50      23      11
>> u=v(3:7)                                A vector u is created from the ele-
u =                                        ments 3 through 7 of vector v.
        8      12      34       2      50
>>
```

For a matrix:

$A(:,n)$ Refers to the elements in all the rows of column n of the matrix A.

$A(n,:)$ Refers to the elements in all the columns of row n of the matrix A.

$A(:,m:n)$ Refers to the elements in all the rows between columns m and n of the matrix A.

$A(m:n,:)$ Refers to the elements in all the columns between rows m and n of the matrix A.

A($m:n,p:q$)   Refers to the elements in rows m through n and columns p through q of the matrix A.

The use of the colon symbol in addressing elements of matrices is demonstrated in Tutorial 2-3.

Tutorial 2-3: Using a colon in addressing arrays.

```
>> A=[1 3 5 7 9 11; 2 4 6 8 10 12; 3 6 9 12 15 18; 4 8 12 16
20 24; 5 10 15 20 25 30]
                                              Define a matrix A with
                                              5 rows and 6 columns.
A =
        1       3       5       7       9      11
        2       4       6       8      10      12
        3       6       9      12      15      18
        4       8      12      16      20      24
        5      10      15      20      25      30
>> B=A(:,3)
                                     Define a column vector B from the
B =                                  elements in all the rows of column 3
        5                            in matrix A.
        6
        9
       12
       15
>> C=A(2,:)                          Define a row vector C from the
                                     elements in all the columns of
                                     row 2 in matrix A.
```

Tutorial 2-3: Using a colon in addressing arrays. (Continued)

```
C =
     2     4     6     8    10    12
>> E=A(2:4,:)
E =
     2     4     6     8    10    12
     3     6     9    12    15    18
     4     8    12    16    20    24
>> F=A(1:3,2:4)
F =
     3     5     7
     4     6     8
     6     9    12
>>
```

Define a matrix E from the elements in rows 2 through 4 and all the columns in matrix A.

Create a matrix F from the elements in rows 1 through 3 and columns 2 through 4 in matrix A.

In Tutorial 2-3 new vectors and matrices were created from existing ones by using a range of elements, or a range of rows and columns (using :). It is possible, however, to select only specific elements, or specific rows and columns of existing variables to create new variables. This is done by typing the selected elements or rows or columns inside brackets, as shown below:

```
>> v=4:3:34
v =
     4     7    10    13    16    19    22    25    28    31    34
>> u=v([3,   5,   7:10])

u =
    10    16    22    25    28    31
>> A=[10:-1:4; ones(1,7); 2:2:14; zeros(1,7)]
A =
    10     9     8     7     6     5     4
     1     1     1     1     1     1     1
     2     4     6     8    10    12    14
     0     0     0     0     0     0     0
>> B = A([1,3],[1,3,5:7])
B =
    10     8     6     5     4
     2     6    10    12    14
```

Create a vector v with 11 elements.

Create a vector u from the 3rd, the 5th, and 7th through 10th elements of v.

Create a 4 × 7 matrix A.

Create a matrix B from the 1st and 3rd rows, and 1st, 3rd, and 5th through 7th columns of A.

2.7 ADDING ELEMENTS TO EXISTING VARIABLES

A variable that exists as a vector, or a matrix, can be changed by adding elements to it (remember that a scalar is a vector with one element). A vector (a matrix with a single row or column) can be changed to have more elements, or it can be changed to be a two-dimensional matrix. Rows and/or columns can also be added to an existing matrix to obtain a matrix of different size. The addition of elements can be done by simply assigning values to the additional elements, or by appending existing variables.

Adding elements to a vector:

Elements can be added to an existing vector by assigning values to the new elements. For example, if a vector has 4 elements, the vector can be made longer by assigning values to elements 5, 6, and so on. If a vector has n elements and a new value is assigned to an element with address of $n + 2$ or larger, MATLAB assigns zeros to the elements that are between the last original element and the new element. Examples:

```
>> DF=1:4                              Define vector DF with 4 elements.
DF =
     1     2     3     4
>> DF(5:10)=10:5:35             Adding 6 elements starting with the 5th.
DF =
     1     2     3     4    10    15    20    25    30    35
>> AD=[5   7   2]                      Define vector AD with 3 elements.
AD =
     5     7     2
>> AD(8)=4                             Assign a value to the 8th element.
AD =                                   MATLAB assigns zeros to
     5     7     2     0     0     0     0     4    the 4th through 7th elements.
>> AR(5)=24                    Assign a value to the 5th element of a new vector.
AR =                                   MATLAB assigns zeros to the
     0     0     0     0    24         1st through 4th elements.
>>
```

Elements can also be added to a vector by appending existing vectors. Two examples are:

```
>> RE=[3   8   1   24];                Define vector RE with 4 elements.
>> GT=4:3:16;                          Define vector GT with 5 elements.
>> KNH=[RE   GT]                       Define a new vector KNH by
KNH =                                  appending RE and GT.
     3     8     1    24     4     7    10    13    16
```

```
>> KNV=[RE'; GT']
KNV =
     3
     8
     1
    24
     4
     7
    10
    13
    16
```

Create a new column vector KNV by appending RE′ and GT′.

Adding elements to a matrix:

Rows and/or columns can be added to an existing matrix by assigning values to the new rows or columns. This can be done by assigning new values, or by appending existing variables. This must be done carefully since the size of the added rows or columns must fit the existing matrix. Examples are:

```
>> E=[1 2 3 4; 5 6 7 8]
E =
     1     2     3     4
     5     6     7     8
```

Define a 2 × 4 matrix E.

```
>> E(3,:)=[10:4:22]

E =
     1     2     3     4
     5     6     7     8
    10    14    18    22
```

Add the vector 10 14 18 22 as the third row of E.

```
>> K=eye(3)
K =
     1     0     0
     0     1     0
     0     0     1
```

Define 3 × 3 matrix K.

```
>> G=[E   K]
G =
     1     2     3     4     1     0     0
     5     6     7     8     0     1     0
    10    14    18    22     0     0     1
```

Append the matrix K to matrix E. The number of rows in E and K must be the same.

If a matrix has a size of $m \times n$, and a new value is assigned to an element with an address beyond the size of the matrix, MATLAB increases the size of the matrix to include the new element. Zeros are assigned to the other elements that are added. Examples:

```
>> AW=[3 6 9; 8 5 11]
AW =
     3     6     9
     8     5    11
```
Define a 2 × 3 matrix.

```
>> AW(4,5)=17
AW =
     3     6     9     0     0
     8     5    11     0     0
     0     0     0     0     0
     0     0     0     0    17
```
Assign a value to the (4,5) element.

MATLAB changes the matrix size to 4 × 5, and assigns zeros to the new elements.

```
>> BG(3,4)=15
```
Assign a value to the (3,4) element of a new matrix.

```
BG =
     0     0     0     0
     0     0     0     0
     0     0     0    15
>>
```

MATLAB creates a 3 × 4 matrix and assigns zeros to all the elements except BG(3,4).

2.8 DELETING ELEMENTS

An element, or a range of elements, of an existing variable can be deleted by reassigning nothing to these elements. This is done by using square brackets with nothing typed in between them. By deleting elements a vector can be made shorter and a matrix can be made to have a smaller size. Examples are:

```
>> kt=[2 8 40 65 3 55 23 15 75 80]
kt =
     2     8    40    65     3    55    23    15    75    80
```
Define a vector with 10 elements.

```
>> kt(6)=[]
```
Eliminate the sixth element.

```
kt =
     2     8    40    65     3    23    15    75    80
```
The vector now has 9 elements.

```
>> kt(3:6)=[]
```
Eliminate elements 3 through 6.

```
kt =
     2     8    15    75    80
```
The vector now has 5 elements.

```
>> mtr=[5 78 4 24 9; 4 0 36 60 12; 56 13 5 89 3]
mtr =
     5    78     4    24     9
     4     0    36    60    12
    56    13     5    89     3
```
Define a 3 × 5 matrix.

```
>> mtr(:,2:4)=[]
```
Eliminate all the rows of columns 2 through 4.

```
mtr =
         5         9
         4        12
        56         3
>>
```

2.9 BUILT-IN FUNCTIONS FOR HANDLING ARRAYS

MATLAB has many built-in functions for managing and handling arrays. Some of these are listed below:

Table 2-2: Built-in functions for handling arrays

Function	Description	Example
length(A)	Returns the number of elements in the vector A.	>> A=[5 9 2 4]; >> length(A) ans = 4

size(A)	Returns a row vector [m,n], where m and n are the size $m \times n$ of the array A.	>> A=[6 1 4 0 12; 5 19 6 8 2]
		A =
		    6 1 4 0 12
		    5 19 6 8 2
		>> size(A)
		ans =
		    2 5

reshape(A, m,n)	Rearrange a matrix A that has r rows and s columns to have m rows and n columns. r times s must be equal to m times n.	>> A=[5 1 6; 8 0 2]
		A =
		    5 1 6
		    8 0 2
		>> B = reshape(A,3,2)
		B =
		    5 0
		    8 6
		    1 2

diag(v)	When v is a vector, creates a square matrix with the elements of v in the diagonal.	>> v=[7 4 2];
		>> A=diag(v)
		A =
		    7 0 0
		    0 4 0
		    0 0 2

Table 2-2: Built-in functions for handling arrays (Continued)

Function	Description	Example

`diag(A)`	When A is a matrix, creates a vector from the diagonal elements of A.	```>> A=[1 2 3; 4 5 6; 7 8 9]```
		`A =`
		     1 2 3
		     4 5 6
		     7 8 9
		`>> vec=diag(A)`
		`vec =`
		     1
		     5
		     9

Additional built-in functions for manipulation of arrays are described in the Help Window. In this window select "Functions by Category", then "Mathematics", and then "Arrays and Matrices."

Sample Problem 2-1: Create a matrix

Using the ones and zeros commands, create a 4×5 matrix in which the first two rows are 0's and the next two rows are 1's.

Solution

```
>> A(1:2,:)=zeros(2,5)
A =
     0     0     0     0     0
     0     0     0     0     0
```
First, create a 2×5 matrix with zeros.

```
>> A(3:4,:)=ones(2,5)
A =
     0     0     0     0     0
     0     0     0     0     0
     1     1     1     1     1
     1     1     1     1     1
```
Add rows 3 and 4 with ones.

A different solution to the problem is:

```
>> A=[zeros(2,5);ones(2,5)]
A =
     0     0     0     0     0
     0     0     0     0     0
     1     1     1     1     1
     1     1     1     1     1
```
Create a 4×5 matrix from two 2×5 matrices.

Sample Problem 2-2: Create a matrix

Create a 6×6 matrix in which the middle two rows, and the middle two columns are 1's, and the rest are 0's.

Solution

```
>> AR=zeros(6,6)
AR =
     0    0    0    0    0    0
     0    0    0    0    0    0
     0    0    0    0    0    0
     0    0    0    0    0    0
     0    0    0    0    0    0
     0    0    0    0    0    0
```

First, create a 6×6 matrix with zeros.

```
>> AR(3:4,:)=ones(2,6)
AR =
     0    0    0    0    0    0
     0    0    0    0    0    0
     1    1    1    1    1    1
     1    1    1    1    1    1
     0    0    0    0    0    0
     0    0    0    0    0    0
```

Reassign the number 1 to the 3rd and 4th rows.

```
>> AR(:,3:4)=ones(6,2)
AR =
     0    0    1    1    0    0
     0    0    1    1    0    0
     1    1    1    1    1    1
     1    1    1    1    1    1
     0    0    1    1    0    0
     0    0    1    1    0    0
```

Reassign the number 1 to the 3rd and 4th columns.

Sample Problem 2-3: Matrix manipulation

Given are a 5×6 matrix A, a 3×6 matrix B, and a 9 element long vector v.

$$A = \begin{bmatrix} 2 & 5 & 8 & 11 & 14 & 17 \\ 3 & 6 & 9 & 12 & 15 & 18 \\ 4 & 7 & 10 & 13 & 16 & 19 \\ 5 & 8 & 11 & 14 & 17 & 20 \\ 6 & 9 & 12 & 15 & 18 & 21 \end{bmatrix}$$

$$B = \begin{bmatrix} 5 & 10 & 15 & 20 & 25 & 30 \\ 30 & 35 & 40 & 45 & 50 & 55 \\ 55 & 60 & 65 & 70 & 75 & 80 \end{bmatrix}$$

$$v = \begin{bmatrix} 99 & 98 & 97 & 96 & 95 & 94 & 93 & 92 & 91 \end{bmatrix}$$

Create the three arrays in the Command Window, and then, by writing one command, replace the last four columns of the 1st and 3rd rows of A with the first four columns of the first two rows of B, the last four columns of the 4th row of A with the elements 5 through 8 of *v*, and the last four columns of the 5th row of *A* with columns 3 through 5 of the third row of *B*.

Solution

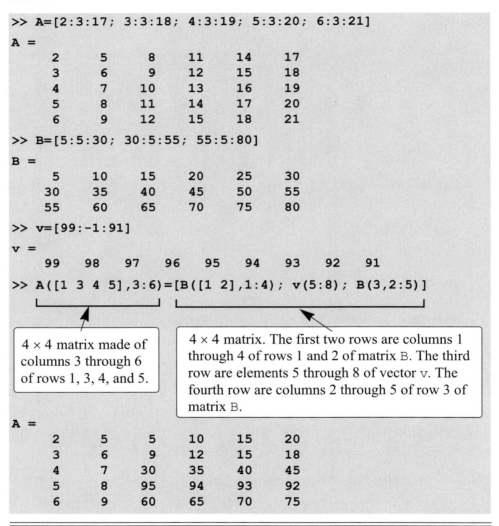

```
>> A=[2:3:17; 3:3:18; 4:3:19; 5:3:20; 6:3:21]

A =
      2      5      8     11     14     17
      3      6      9     12     15     18
      4      7     10     13     16     19
      5      8     11     14     17     20
      6      9     12     15     18     21

>> B=[5:5:30; 30:5:55; 55:5:80]

B =
      5     10     15     20     25     30
     30     35     40     45     50     55
     55     60     65     70     75     80

>> v=[99:-1:91]

v =
     99     98     97     96     95     94     93     92     91

>> A([1 3 4 5],3:6)=[B([1 2],1:4); v(5:8); B(3,2:5)]
```

4 × 4 matrix made of columns 3 through 6 of rows 1, 3, 4, and 5.

4 × 4 matrix. The first two rows are columns 1 through 4 of rows 1 and 2 of matrix B. The third row are elements 5 through 8 of vector v. The fourth row are columns 2 through 5 of row 3 of matrix B.

```
A =
      2      5      5     10     15     20
      3      6      9     12     15     18
      4      7     30     35     40     45
      5      8     95     94     93     92
      6      9     60     65     70     75
```

2.10 STRINGS AND STRINGS AS VARIABLES

• A string is an array of characters. It is created by typing the characters within single quotes.

• Strings can include letters, digits, other symbols, and spaces.

• Examples of strings: 'ad ef ', '3%fr2', '{edcba:21!', 'MATLAB'.

- A string that contains a single quote is created by typing two single quotes within the string.

- When a string is being typed in, the color of the text on the screen changes to maroon when the first single quote is typed. When the single quote at the end of the string is typed the color of the string changes to purple.

Strings have several different uses in MATLAB. They are used in output commands to display text messages (Chapter 4), in formatting commands of plots (Chapter 5), and as input arguments of some functions (Chapter 6). More details are given in the chapters when strings are used for these purposes.

- When strings are being used in formatting plots (labels to axes, title, and text notes), characters within the string can be formatted to have a specified font, size, position (uppercase, lowercase), color, etc. See Chapter 5 for details.

Strings can also be assigned to variables by simply typing the string on the right side of the assignment operator, as shown in the examples below:

```
>> a='FRty 8'
a =
FRty 8
>> B='My name is John Smith'
B =
My name is John Smith
>>
```

When a variable is defined as a string, the characters of the string are stored in an array just as numbers are. Each character, including a space, is an element in the array. This means that a one-line string is a row vector in which the number of elements is equal to the number of characters. The elements of the vectors are addressed by their position. For example, in the vector B that was defined above the fourth element is the letter n, the twelfth element is J and so on.

```
>> B(4)
ans =
n
>> B(12)
ans =
J
>>
```

As with a vector that contains numbers, it is also possible to change specific elements by addressing them directly. For example, in the vector B above the name John can be changed to Bill by:

```
>> B(12:15)='Bill'

B =

My name is Bill Smith

>>
```

> Using a colon to assign new characters to elements 12 through 15 in the vector B.

Strings can also be placed in a matrix. As with numbers, this is done by typing a semicolon ; (or pressing the **Enter** key) at the end of each row. Each row must be typed as a string, which means that it must be enclosed in single quotes. In addition, as with a numerical matrix, the number of elements in all the rows must be the same. This requirement can cause problems when the intention is to create rows with specific wording. Rows can be made to have the same number of elements by adding spaces.

MATLAB has a built-in function named char that creates an array with rows that have the same number of characters from an input of rows that are not of the same length. MATLAB makes the length of all the rows equal to the longest row by adding spaces at the end of the short lines. In the char function, the rows are entered as strings separated by a comma according to the following format:

```
variable_ name = char('string 1','string 2','string 3')
```

For example:

```
>> Info=char('Student Name:','John Smith','Grade:','A+')

Info =
Student Name:
John Smith
Grade:
A+
>>
```

> A variable named Info is assigned four rows of strings, each with different length.

> The function char creates an array with four rows with the same length as the longest row by adding empty spaces to the shorter lines.

A variable can be defined as a number or a string that is made up of the same digits. For example, as shown below, x is defined to be the number 536, and y is defined to be a string made up of the digits 536.

```
>> x=536

x =
   536
>> y='536'

y =
536
>>
```

The two variables are not the same even though they appear identical on the screen. The variable x can be used in mathematical expressions, while the variable y can not.

2.11 PROBLEMS

1. Create a row vector that has the elements: 6, $8 \cdot 3$, 81, $e^{2.5}$, $\sqrt{65}$, $\sin(\pi/3)$, and 23.05.

2. Create a column vector that has the elements: 44, 9, $\ln(51)$, 2^3, 0.1, and $5\tan(25°)$.

3. Create a row vector in which the first element is 0, and the last element is 42, with an increment of 3 between the elements (0, 3, 6, ..., 42).

4. Create a column vector in which the first element is 18, the elements decrease with increments of –4, and the last element is –22. (A column vector can be created by the transpose of a row vector.)

5. Create a row vector with 16 equally spaced elements in which the first element is 5 and the last element is 61.

6. Create a column vector with 14 equally spaced elements in which the first element is 3 and the last element is –36.

7. Using the colon symbol, create a row vector (assign it to a variable named `same`) with eleven elements that are all 4.

8. Create a vector, name it `Afirst`, that has 13 elements in which the first is 3, the increment is 4 and the last element is 51. Then, using the colon symbol, create a new vector, call it `Asecond`, that has seven elements. The first four elements are the first four elements of the vector `Afirst`, and the last three are the last three elements of the vector `Afirst`.

9. Create the matrix shown below by using the vector notation for creating vectors with constant spacing and/or the `linspace` command when entering the rows.

$$B = \begin{bmatrix} 0 & 4 & 8 & 12 & 16 & 20 & 24 & 28 \\ 69 & 68 & 67 & 66 & 65 & 64 & 63 & 62 \\ 1.4 & 1.1 & 0.8 & 0.5 & 0.2 & -0.1 & -0.4 & -0.7 \end{bmatrix}$$

10. Using the colon symbol, create a 3×5 matrix (assign it to a variable named `msame`) in which all the elements are the number 7.

11. Create three vectors:
$$a = \begin{bmatrix} 2 & -1 & 0 & 6 \end{bmatrix}, \quad b = \begin{bmatrix} -5 & 20 & 12 & -3 \end{bmatrix}, \quad c = \begin{bmatrix} 10 & 7 & -2 & 1 \end{bmatrix}$$

   a) Use the three vectors in a MATLAB command to create a 3×4 matrix in which the rows are the vectors a, b, and c.

   b) Use the three vectors in a MATLAB command to create a 4×3 matrix in which the columns are the vectors a, b, and c.

12. By hand (pencil and paper) write what will be displayed if the following commands are executed by MATLAB. Check your answers by executing the commands with MATLAB. (Parts b, c, and d use the vector that was defined in part a.)

   a) `a=0:2:6`   b) `b=[a a]` or `b=[a,a]`

   c) `c=[a; a]`

   d) `d=[a' a']` or `d=[a',a']`

13. The following vector is defined in MATLAB:
$$v = [2, 7, -3, 5, 0, 14, -1, 10, -6, 8]$$
   By hand (pencil and paper) write what will be displayed if the following commands are executed by MATLAB. Check your answers by executing the commands with MATLAB.

   a) `a=v(3:6)`   b) `b=v([2,4:7,10])`   c) `c=v([9,3,1,10])`

   d) `d=[v([1,3 5]);v([2,4,6]);v([3,6,9])]`

14. Create the following matrix A:
$$A = \begin{bmatrix} 1 & 2 & 3 & 4 & 5 \\ 6 & 7 & 8 & 9 & 10 \\ 11 & 12 & 13 & 14 & 15 \end{bmatrix}$$

   Use the matrix A to:

   a) Create a five-element row vector named `va` that contains the elements of the first row of A.

   b) Create a three-element row vector named `vb` that contains the elements of the third column of A.

   c) Create a eight-element row vector named `vc` that contains the elements of the second rows of A and the fourth column of A.

   d) Create a six-element row vector named `vd` that contains the elements of the first and fifth columns of A.

15. Create the following matrix B:
$$B = \begin{bmatrix} 15 & 12 & 9 & 6 & 3 \\ 2 & 4 & 6 & 8 & 10 \\ 6 & 12 & 18 & 24 & 30 \end{bmatrix}$$

   Use the matrix B to:

   a) Create a six-element column vector named `ua` that contains the elements of the second and fourth columns of C.

b) Create a five-element column vector named ub that contains the elements of the third row of C.

c) Create a nine-element row vector named uc that contains the elements of the second, fourth and fifth columns of C.

d) Create a eight-element row vector named ud that contains the elements of the first column and first row of C.

16. Create the following matrix A:

$$\begin{bmatrix} 0.1 & 0.2 & 0.3 & 0.4 & 0.5 & 0.6 & 0.7 \\ 14 & 12 & 10 & 8 & 6 & 4 & 2 \\ 1 & 1 & 1 & 1 & 0 & 0 & 0 \\ 3 & 6 & 9 & 12 & 15 & 18 & 21 \end{bmatrix}$$

a) Create a 3×4 matrix B from the 1st, 2nd, and 3rd rows, and the 1st through the 4th columns of the matrix A.

b) Create a 2×7 matrix C from the 2nd, and 3rd rows, and all the columns of the matrix A.

17. The following matrix is defined in MATLAB: $M =$
$$\begin{bmatrix} 6 & 9 & 12 & 15 & 18 & 21 \\ 4 & 4 & 4 & 4 & 4 & 4 \\ 2 & 1 & 0 & -1 & -2 & -3 \\ -6 & -4 & -2 & 0 & 2 & 4 \end{bmatrix}$$

By hand (pencil and paper) write what will be displayed if the following commands are executed by MATLAB. Check your answers by executing the commands with MATLAB.

a) A=M([1,3],[2,4])

b) B=M(:,[1,4:6])

c) C=M([2,3],:)

18. Using the zeros, ones, and eye commands create the following arrays:

a) $\begin{bmatrix} 0 & 0 & 0 & 1 & 1 & 1 \\ 0 & 0 & 0 & 1 & 1 & 1 \end{bmatrix}$

b) $\begin{bmatrix} 1 & 1 & 0 & 0 & 0 & 0 \\ 1 & 0 & 1 & 0 & 0 & 0 \\ 1 & 0 & 0 & 1 & 0 & 0 \\ 1 & 0 & 0 & 0 & 1 & 0 \end{bmatrix}$

c) $\begin{bmatrix} 1 & 1 \\ 0 & 0 \\ 0 & 0 \\ 1 & 1 \end{bmatrix}$

19. Use the `eye` command to create the array A shown on the left below, Then, using the colon to address the elements in the arrays, change A to be like the one shown on the right.

$$A = \begin{bmatrix} 1 & 0 & 0 & 0 & 0 & 0 \\ 0 & 1 & 0 & 0 & 0 & 0 \\ 0 & 0 & 1 & 0 & 0 & 0 \\ 0 & 0 & 0 & 1 & 0 & 0 \\ 0 & 0 & 0 & 0 & 1 & 0 \\ 0 & 0 & 0 & 0 & 0 & 1 \end{bmatrix} \qquad A = \begin{bmatrix} 1 & 0 & 0 & 3 & 3 & 3 \\ 0 & 1 & 0 & 3 & 3 & 3 \\ 0 & 0 & 1 & 3 & 3 & 3 \\ 0 & 0 & 0 & 1 & 0 & 0 \\ 2 & 2 & 2 & 2 & 1 & 0 \\ 2 & 2 & 2 & 2 & 0 & 1 \end{bmatrix}$$

20. Create a vector $v = [1, 2, 3, ..., 35]$ with 35 elements. Then, use the reshape function to create a 5×7 matrix in which the first row is the numbers 1 2 3 4 5 6 7, the second row is the numbers 8 9 10 11 12 13 14, the third row is the numbers 15 through 21, and so on.

21. Create a 3×3 matrix A in which all the elements are 1. Then, reassign A to itself such that A will become:

$$A = \begin{bmatrix} 1 & 1 & 1 & 0 & 0 & 0 \\ 1 & 1 & 1 & 0 & 0 & 0 \\ 1 & 1 & 1 & 0 & 0 & 0 \\ 0 & 0 & 0 & 1 & 1 & 1 \\ 0 & 0 & 0 & 1 & 1 & 1 \\ 0 & 0 & 0 & 1 & 1 & 1 \end{bmatrix}$$

Chapter 3

Mathematical Operations with Arrays

Once variables are created in MATLAB they can be used in a wide variety of mathematical operations. In Chapter 1 the variables that were used in mathematical operations were all defined as scalars. This means that they were all 1×1 arrays (arrays with one row and one column that have only one element) and the mathematical operations were done with single numbers. Arrays, however, can be one-dimensional (arrays with one row, or with one column), two-dimensional (arrays with rows and columns), and even of higher dimensions. In these cases the mathematical operations are more complex. MATLAB, as its name indicates, is designed to carry out advanced array operations that have many applications in science and engineering. This chapter presents the basic, most common mathematical operations that MATLAB performs using arrays.

Addition and subtraction are relatively simple operations and are covered first in Section 3.1. The other basic operations, multiplication, division and exponentiation, can be done in MATLAB in two different ways. One way, which uses the standard symbols (*, /, and ^), follows the rules of linear algebra and is presented in Sections 3.2 and 3.3. The second way, which is called element-by-element operations, is covered in Section 3.4. These operations use the symbols .*, ./, and .^ (a period is typed in front of the standard operation symbol). In addition, in both types of calculations, MATLAB has left division operators (.\ or \ ), which are also explained in Sections 3.3 and 3.4.

A note to first time users of MATLAB:

Although matrix operations are presented first and element-by-element operations next, the order can be reversed since the two are independent of each other. It is expected that almost every user has some knowledge of matrix operations and linear algebra, and thus will be able to follow the material covered in Sections 3.2 and 3.3 without any difficulty. Some readers, however, might prefer to read Section 3.4 first. MATLAB can be used with element-by-element operations in numerous applications that do not require linear algebra multiplication (or division) operations.

3.1 ADDITION AND SUBTRACTION

The operations + (addition) and − (subtraction) can be used to add (subtract) arrays of identical size (the same number of rows and columns), and to add (subtract) a scalar to an array. When two arrays are involved, the sum, or the difference of two arrays is obtained by adding, or subtracting, their corresponding elements.

In general, if A and B are two arrays (for example 2×3 matrices),

$$A = \begin{bmatrix} A_{11} & A_{12} & A_{13} \\ A_{21} & A_{22} & A_{23} \end{bmatrix} \quad \text{and} \quad B = \begin{bmatrix} B_{11} & B_{12} & B_{13} \\ B_{21} & B_{22} & B_{23} \end{bmatrix}$$

then, the matrix that is obtained by adding A and B is:

$$\begin{bmatrix} (A_{11} + B_{11}) & (A_{12} + B_{12}) & (A_{13} + B_{13}) \\ (A_{21} + B_{21}) & (A_{22} + B_{22}) & (A_{23} + B_{23}) \end{bmatrix}$$

When a scalar (number) is added to (or subtracted from) an array, the number is added to (or subtracted from) all the elements of the array. Examples are:

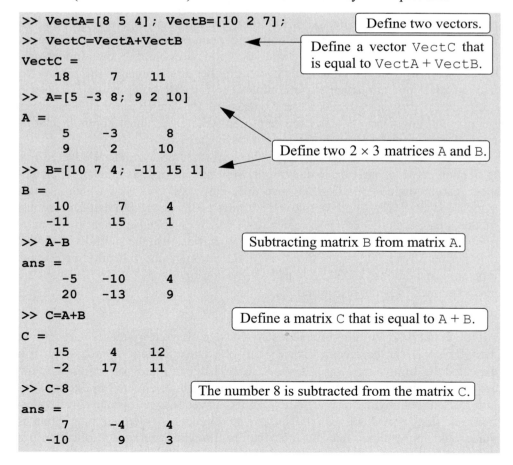

```
>> VectA=[8 5 4]; VectB=[10 2 7];              Define two vectors.

>> VectC=VectA+VectB          ◄————    Define a vector VectC that
                                        is equal to VectA + VectB.
VectC =
     18        7       11
>> A=[5 -3 8; 9 2 10]
A =
      5       -3        8
      9        2       10                Define two 2 × 3 matrices A and B.
>> B=[10 7 4; -11 15 1]
B =
     10        7        4
    -11       15        1
>> A-B                               Subtracting matrix B from matrix A.
ans =
     -5      -10        4
     20      -13        9
>> C=A+B                         Define a matrix C that is equal to A + B.
C =
     15        4       12
     -2       17       11
>> C-8                         The number 8 is subtracted from the matrix C.
ans =
      7       -4        4
    -10        9        3
```

When a scalar and an array are involved in the addition (subtraction) operation, the scalar is added to (or subtracted from) all the elements in the matrix. Examples are:

```
>> VectA=[1 5 8 -10 2]          Define a vector named VectA.
VectA =
       1       5       8     -10       2
>> VectA+4                       Add the scalar 4 to VectA.
ans =
       5       9      12      -6        4 is added to each element of VectA.
>> A=[6 21 -15; 0 -4 8]          Define a 2 × 3 matrix A.
A =
       6      21     -15
       0      -4       8
>> A-5                           Subtract the scalar 5 from A.
ans =
       1      16     -20
      -5      -9       3         5 is subtracted from each element of A.
```

3.2 ARRAY MULTIPLICATION

The multiplication operation * is executed by MATLAB according to the rules of linear algebra. This means that if A and B are two matrices, the operation $A*B$ can be carried out only if the number of columns in matrix A is equal to the number of rows in matrix B. The result is a matrix that has the same number of rows as A and the same number of columns as B. For example, if A is a 4×3 matrix and B is a 3×2 matrix:

$$A = \begin{bmatrix} A_{11} & A_{12} & A_{13} \\ A_{21} & A_{22} & A_{23} \\ A_{31} & A_{32} & A_{33} \\ A_{41} & A_{42} & A_{43} \end{bmatrix} \text{ and } B = \begin{bmatrix} B_{11} & B_{12} \\ B_{21} & B_{22} \\ B_{31} & B_{32} \end{bmatrix}$$

then, the matrix that is obtained by the operation $A*B$ has the dimension of 4×2 with the elements:

$$\begin{bmatrix} (A_{11}B_{11}+A_{12}B_{21}+A_{13}B_{31}) & (A_{11}B_{12}+A_{12}B_{22}+A_{13}B_{32}) \\ (A_{21}B_{11}+A_{22}B_{21}+A_{23}B_{31}) & (A_{21}B_{12}+A_{22}B_{22}+A_{23}B_{32}) \\ (A_{31}B_{11}+A_{32}B_{21}+A_{33}B_{31}) & (A_{31}B_{12}+A_{32}B_{22}+A_{33}B_{32}) \\ (A_{41}B_{11}+A_{42}B_{21}+A_{43}B_{31}) & (A_{41}B_{12}+A_{42}B_{22}+A_{43}B_{32}) \end{bmatrix}$$

A numerical example is:

$$\begin{bmatrix} 1 & 4 & 3 \\ 2 & 6 & 1 \\ 5 & 2 & 8 \end{bmatrix} \begin{bmatrix} 5 & 4 \\ 1 & 3 \\ 2 & 6 \end{bmatrix} = \begin{bmatrix} (1\cdot5+4\cdot1+3\cdot2) & (1\cdot4+4\cdot3+3\cdot6) \\ (2\cdot5+6\cdot1+1\cdot2) & (2\cdot4+6\cdot3+1\cdot6) \\ (5\cdot5+2\cdot1+8\cdot2) & (5\cdot4+2\cdot3+8\cdot6) \end{bmatrix} = \begin{bmatrix} 15 & 34 \\ 18 & 32 \\ 43 & 74 \end{bmatrix}$$

The product of the multiplication of two square matrices (they must be of the same size) is also a square matrix of the same size. However, the multiplication of matrices is not commutative. This means that if A and B are both $n \times n$, then $A*B \neq B*A$. Also, the power operation can only be executed with a square matrix (since $A*A$ can be carried out only if the number of columns in the first matrix is equal to the number of rows in the second matrix).

Two vectors can multiply each other only if both have the same number of elements, and one is a row vector and the other is a column vector. The multiplication of a row vector times a column vector gives a 1×1 matrix, which is a scalar. This is the dot product of two vectors. (MATLAB also has a built-in function, named dot(a,b), that computes the dot product of two vectors.) When using the dot function, the vectors a and b can each be a row or a column vector, (see Table 3-1). The multiplication of a column vector times a row vector, both with n elements gives an $n \times n$ matrix.

Tutorial 3-1: Multiplication of arrays.

```
>> A=[1 4 2; 5 7 3; 9 1 6; 4 2 8]

A =
        1      4      2
        5      7      3
        9      1      6
        4      2      8
```
Define a 4 × 3 matrix A.

```
>> B=[6 1; 2 5; 7 3]

B =
        6      1
        2      5
        7      3
```
Define a 3 × 2 matrix B.

```
>> C=A*B
```
Multiply matrix A by matrix B and assign the result to variable C.

```
C =
       28     27
       65     49
       98     32
       84     38
```

```
>> D=B*A
??? Error using ==> *
Inner matrix dimensions must agree.
```
Trying to multiply B by A, B*A, gives an error since the number of columns in B is 2, and the number of rows in A is 4.

```
>> F=[1 3; 5 7]

F =
        1      3
        5      7
```
Define two 2 × 2 matrices F and G.

Tutorial 3-1: Multiplication of arrays. (Continued)

```
>> G=[4 2; 1 6]
G =
     4      2
     1      6
>> F*G                                          Multiply F*G
ans =
     7     20
    27     52
>> G*F                                          Multiply G*F
ans =
    14     26         Note that the answer for G*F is not the
    31     45         same as the answer for F*G.
>> AV=[2 5 1]        Define a three-element row vector AV.
AV =
     2      5      1
>> BV=[3; 1; 4]       Define a three-element column vector BV.
BV =
     3
     1
     4
>> AV*BV              Multiply AV by BV. The answer is a scalar.
ans =                 (Dot product of two vectors.)
    15
>> BV*AV              Multiply BV by AV. The
ans =                 answer is a 3 × 3 matrix.
     6     15      3
     2      5      1
     8     20      4
>>
```

When an array is multiplied by a number (actually a number is a 1×1 array), each element in the array is multiplied by the number. For example:

```
>> A=[2 5 7 0; 10 1 3 4; 6 2 11 5]     Define a 3 × 4 matrix A.
A =
     2      5      7      0
    10      1      3      4
     6      2     11      5
>> b=3                              Assign the number 3 to the variable b.
```

```
b =
     3
>> b*A
```

> Multiply the matrix A by b. This can be done by either typing b*A or A*b.

```
ans =
     6      15      21       0
    30       3       9      12
    18       6      33      15
>> C=A*5

C =
    10      25      35       0
    50       5      15      20
    30      10      55      25
```

> Multiply the matrix A by 5 and assign the result to a new variable C. (Typing C = 5*A gives the same result.)

Linear algebra rules of array multiplication provide a convenient way for writing a system of linear equations. For example, the following system of three equations with three unknowns:

$$A_{11}x_1 + A_{12}x_2 + A_{13}x_3 = B_1$$
$$A_{21}x_1 + A_{22}x_2 + A_{23}x_3 = B_2$$
$$A_{31}x_1 + A_{32}x_2 + A_{33}x_3 = B_3$$

can be written in a matrix form by:

$$\begin{bmatrix} A_{11} & A_{12} & A_{13} \\ A_{21} & A_{22} & A_{23} \\ A_{31} & A_{32} & A_{33} \end{bmatrix} \begin{bmatrix} x_1 \\ x_2 \\ x_3 \end{bmatrix} = \begin{bmatrix} B_1 \\ B_2 \\ B_3 \end{bmatrix}$$

and in matrix notation by:

$$AX = B \quad \text{where } A = \begin{bmatrix} A_{11} & A_{12} & A_{13} \\ A_{21} & A_{22} & A_{23} \\ A_{31} & A_{32} & A_{33} \end{bmatrix}, X = \begin{bmatrix} x_1 \\ x_2 \\ x_3 \end{bmatrix}, \text{ and } B = \begin{bmatrix} B_1 \\ B_2 \\ B_3 \end{bmatrix}.$$

3.3 ARRAY DIVISION

The division operation is also associated with the rules of linear algebra. This operation is more complex and only a brief explanation is given below. A full explanation can be found in books on linear algebra.

The division operation can be explained with the help of the identity matrix and the inverse operation.

Identity matrix:

The identity matrix is a square matrix in which the diagonal elements are 1's, and the rest of the elements are 0's. As was shown in Section 2.2.1, an identity matrix can be created in MATLAB with the `eye` command. When the identity matrix multiplies another matrix (or vector), that matrix (or vector) is unchanged (the multiplication has to be done according to the rules of linear algebra). This is equivalent to multiplying a scalar by 1. For example:

$$\begin{bmatrix} 7 & 3 & 8 \\ 4 & 11 & 5 \end{bmatrix} \begin{bmatrix} 1 & 0 & 0 \\ 0 & 1 & 0 \\ 0 & 0 & 1 \end{bmatrix} = \begin{bmatrix} 7 & 3 & 8 \\ 4 & 11 & 5 \end{bmatrix} \text{ or } \begin{bmatrix} 1 & 0 & 0 \\ 0 & 1 & 0 \\ 0 & 0 & 1 \end{bmatrix} \begin{bmatrix} 8 \\ 2 \\ 15 \end{bmatrix} = \begin{bmatrix} 8 \\ 2 \\ 15 \end{bmatrix} \text{ or } \begin{bmatrix} 6 & 2 & 9 \\ 1 & 8 & 3 \\ 7 & 4 & 5 \end{bmatrix} \begin{bmatrix} 1 & 0 & 0 \\ 0 & 1 & 0 \\ 0 & 0 & 1 \end{bmatrix} = \begin{bmatrix} 6 & 2 & 9 \\ 1 & 8 & 3 \\ 7 & 4 & 5 \end{bmatrix}$$

If a matrix A is square, it can be multiplied by the identity matrix, I, from the left or from the right:

$$AI = IA = A$$

Inverse of a matrix:

The matrix B is the inverse of the matrix A if when the two matrices are multiplied the product is the identity matrix. Both matrices must be square and the multiplication order can be BA or AB.

$$BA = AB = I$$

Obviously B is the inverse of A, and A is the inverse of B. For example:

$$\begin{bmatrix} 2 & 1 & 4 \\ 4 & 1 & 8 \\ 2 & -1 & 3 \end{bmatrix} \begin{bmatrix} 5.5 & -3.5 & 2 \\ 2 & -1 & 0 \\ -3 & 2 & 1 \end{bmatrix} = \begin{bmatrix} 5.5 & -3.5 & 2 \\ 2 & -1 & 0 \\ -3 & 2 & 1 \end{bmatrix} \begin{bmatrix} 2 & 1 & 4 \\ 4 & 1 & 8 \\ 2 & -1 & 3 \end{bmatrix} = \begin{bmatrix} 1 & 0 & 0 \\ 0 & 1 & 0 \\ 0 & 0 & 1 \end{bmatrix}$$

The inverse of a matrix A is typically written as A^{-1}. In MATLAB the inverse of a matrix can be obtained either by raising A to the power of -1, A^{-1}, or with the `inv(A)` function. For example, multiplying the matrices above with MATLAB is shown below:

```
>> A=[2 1 4; 4 1 8; 2 -1 3]
A =
     2     1     4
     4     1     8
     2    -1     3
```
| Creating the matrix A. |

```
>> B=inv(A)
B =
    5.5000   -3.5000    2.0000
    2.0000   -1.0000         0
   -3.0000    2.0000   -1.0000
```
| Use the `inv` function to find the inverse of A and assign it to B. |

```
>> A*B
ans =
     1     0     0
     0     1     0
     0     0     1
```
| Multiplication of A and B gives the identity matrix. |

```
>> A*A^-1

ans =
      1      0      0
      0      1      0
      0      0      1
```

Use power of –1 to find the inverse of A. Multiplying it by A gives the identity matrix.

- Not every matrix has an inverse. A matrix has an inverse only if it is square and its determinant is not equal to zero.

Determinants:

Determinant is a function associated with square matrices. A short review on determinants is given below. For a more detailed coverage the reader is referred to books on linear algebra.

The determinant is a function that associates with each square matrix A a number, called the determinant of the matrix. The determinant is typically denoted by $\det(A)$ or $|A|$. The determinant is calculated according to specific rules. For a second order 2×2 matrix the rule is:

$$A = \begin{vmatrix} a_{11} & a_{12} \\ a_{21} & a_{22} \end{vmatrix} = a_{11}a_{22} - a_{12}a_{21}, \text{ for example, } \begin{vmatrix} 6 & 5 \\ 3 & 9 \end{vmatrix} = 6 \cdot 9 - 5 \cdot 3 = 39$$

The determinant of a square matrix can be calculated with the $\det$ command (see Table 3-1).

Array division:

MATLAB has two types of array division, which are the right division and the left division.

Left division \ :

The left division is used to solve the matrix equation $AX = B$. In this equation X and B are column vectors. This equation can be solved by multiplying on the left both sides by the inverse of A:

$$A^{-1}AX = A^{-1}B$$

The left-hand side of this equation is X since:

$$A^{-1}AX = IX = X$$

So, the solution of $AX = B$ is:

$$X = A^{-1}B$$

In MATLAB the last equation can be written by using the left division character:

$$X = A \backslash B$$

It should be pointed out here that although the last two operations appear to give the same result, the method by which MATLAB calculates X is different. In the first, MATLAB calculates A^{-1} and then use it to multiply B. In the second, (left division) the solution X is obtained numerically with a method that is based on the Gauss elimination method. The left division method is recommended for solving a set of linear equations because the calculation of the inverse may be less accurate than the Gauss elimination method when large matrices are involved.

Right Division / :

The right division is used to solve the matrix equation $XC = D$. In this equation X and D are row vectors. This equation can be solved by multiplying on the right both sides by the inverse of C:

$$X \cdot CC^{-1} = D \cdot C^{-1}$$

which gives:

$$X = D \cdot C^{-1}$$

In MATLAB the last equation can be written by using the right division character:

$$X = D/C$$

The following example demonstrates the use of the left and right divisions, and the `inv` function to solve a set of linear equations.

Sample Problem 3-1: Solving three linear equations (array division)

Use matrix operations to solve the following system of linear equations.

$$4x - 2y + 6z = 8$$
$$2x + 8y + 2z = 4$$
$$6x + 10y + 3z = 0$$

Solution

Using the rules of linear algebra demonstrated earlier, the above system of equations can be written in the matrix form $AX = B$ or in the form $XC = D$:

$$\begin{bmatrix} 4 & -2 & 6 \\ 2 & 8 & 2 \\ 6 & 10 & 3 \end{bmatrix} \begin{bmatrix} x \\ y \\ z \end{bmatrix} = \begin{bmatrix} 8 \\ 4 \\ 0 \end{bmatrix} \quad \text{or} \quad \begin{bmatrix} x & y & z \end{bmatrix} \begin{bmatrix} 4 & 2 & 6 \\ -2 & 8 & 10 \\ 6 & 2 & 3 \end{bmatrix} = \begin{bmatrix} 8 & 4 & 0 \end{bmatrix}$$

The solution of both forms is shown below:

```
>> A=[4 -2 6; 2 8 2; 6 10 3];          | Solving the form AX = B.
>> B=[8; 4; 0];
>> X=A\B                               | Solving by using left division X = A \ B.
X =
   -1.8049
    0.2927
    2.6341
>> Xb=inv(A)*B                         | Solving by using the inverse of A  X = A⁻¹B.
```

```
Xb =
   -1.8049
    0.2927
    2.6341
>> C=[4 2 6; -2 8 10; 6 2 3];          Solving the form XC = D.
>> D=[8 4 0];
>> Xc=D/C                              Solving by using right division X = D/C.
Xc =
   -1.8049      0.2927      2.6341
>> Xd=D*inv(C)                         Solving by using the inverse of C, X = D · C⁻¹.
Xd =
   -1.8049      0.2927      2.6341
```

The boxed annotations read:
- Solving the form $XC = D$.
- Solving by using right division $X = D/C$.
- Solving by using the inverse of C, $X = D \cdot C^{-1}$.

3.4 ELEMENT-BY-ELEMENT OPERATIONS

In Sections 3.2 and 3.3 it was shown that when the regular symbols for multiplication and division are used with arrays (* and /), the mathematical operations follow the rules of linear algebra. There are, however, many situations that require element-by-element operations. These operations are carried out on each of the elements of the array (or arrays). Addition and subtraction are by definition already element-by-element operations since when two arrays are added (or subtracted) the operation is executed with the elements that are in the same position in the arrays. Element-by-element operations can only be done with arrays of the same size.

Element-by-element multiplication, division, and exponentiation of two vectors or matrices is entered in MATLAB by typing a period in front of the arithmetic operator.

Symbol	Description		Symbol	Description
.*	Multiplication		./	Right division
.^	Exponentiation		.\	Left Division

If two vectors a and b are: $a = \begin{bmatrix} a_1 & a_2 & a_3 & a_4 \end{bmatrix}$ and $b = \begin{bmatrix} b_1 & b_2 & b_3 & b_4 \end{bmatrix}$, then element-by-element multiplication, division, and exponentiation of the two vectors gives:

$$a .* b = \begin{bmatrix} a_1 b_1 & a_2 b_2 & a_3 b_3 & a_4 b_4 \end{bmatrix}$$

$$a ./ b = \begin{bmatrix} a_1/b_1 & a_2/b_2 & a_3/b_3 & a_4/b_4 \end{bmatrix}$$

$$a .^\wedge b = \begin{bmatrix} (a_1)^{b_1} & (a_2)^{b_2} & (a_3)^{b_3} & (a_4)^{b_4} \end{bmatrix}$$

If two matrices A and B are:

$$A = \begin{bmatrix} A_{11} & A_{12} & A_{13} \\ A_{21} & A_{22} & A_{23} \\ A_{31} & A_{32} & A_{33} \end{bmatrix} \text{ and } B = \begin{bmatrix} B_{11} & B_{12} & B_{13} \\ B_{21} & B_{22} & B_{23} \\ B_{31} & B_{32} & B_{33} \end{bmatrix}$$

then element-by-element multiplication and division of the two matrices gives:

$$A .* B = \begin{bmatrix} A_{11}B_{11} & A_{12}B_{12} & A_{13}B_{13} \\ A_{21}B_{21} & A_{22}B_{22} & A_{23}B_{23} \\ A_{31}B_{31} & A_{32}B_{32} & A_{33}B_{33} \end{bmatrix} \qquad A ./ B = \begin{bmatrix} A_{11}/B_{11} & A_{12}/B_{12} & A_{13}/B_{13} \\ A_{21}/B_{21} & A_{22}/B_{22} & A_{23}/B_{23} \\ A_{31}/B_{31} & A_{32}/B_{32} & A_{33}/B_{33} \end{bmatrix}$$

Element-by-element exponentiation of matrix A gives:

$$A .^\wedge n = \begin{bmatrix} (A_{11})^n & (A_{12})^n & (A_{13})^n \\ (A_{21})^n & (A_{22})^n & (A_{23})^n \\ (A_{31})^n & (A_{32})^n & (A_{33})^n \end{bmatrix}$$

Element-by-element multiplication, division, and exponentiation are demonstrated in Tutorial 3-2.

Tutorial 3-2: Element-by-element operations.

```
>> A=[2 6 3; 5 8 4]
A =
     2     6     3
     5     8     4
>> B=[1 4 10; 3 2 7]
B =
     1     4    10
     3     2     7
>> A.*B
ans =
     2    24    30
    15    16    28
>> C=A./B
C =
    2.0000    1.5000    0.3000
    1.6667    4.0000    0.5714
```

Define a 2 × 3 array A.

Define a 2 × 3 array B.

Element-by-element multiplication of array A by B.

Element-by-element division of array A by B. The result is assigned to variable C.

Tutorial 3-2: Element-by-element operations. (Continued)

```
>> B.^3

ans =
       1      64    1000
      27       8     343
```
Element-by-element exponentiation of array B. The result is an array in which each term is the corresponding term in B raised to the power of 3.

```
>> A*B


??? Error using ==> *
Inner matrix dimensions must
```
Trying to multiply A*B gives an error since A and B cannot be multiplied according to linear algebra rules. (The number of columns in A is not equal to the number of rows in B.)

Element-by-element calculations are very useful for calculating the value of a function at many values of its argument. This is done by first defining a vector that contains values of the independent variable, and then using this vector in element-by-element computations to create a vector in which each element is the corresponding value of the function. One example is:

```
>> x=[1:8]
x =
     1    2    3    4    5    6    7    8
>> y=x.^2-4*x
y =
    -3   -4   -3    0    5   12   21   32
>>
```
Create a vector x with eight elements.

Vector x is used in element-by-element calculations of the elements of vector y.

In the example above $y = x^2 - 4x$. Element-by-element operation is needed when x is squared. Each element in the vector y is the value of y that is obtained when the value of the corresponding element of the vector x is substituted in the equation. Another example is:

```
>> z=[1:2:11]
z =
     1    3    5    7    9   11
>> y=(z.^3 + 5*z)./(4*z.^2 - 10)

y =
   -1.0000   1.6154   1.6667   2.0323   2.4650   2.9241
```
Create a vector z with eight elements.

Vector z is used in element-by-element calculations of the elements of vector y.

In the last example $y = \dfrac{z^3 + 5z}{4z^2 - 10}$. Element-by-element operations are used in this example three times; to calculate z^3 and z^2, and to divide the numerator by the denominator.

3.5 USING ARRAYS IN MATLAB BUILT-IN MATH FUNCTIONS

The built-in functions in MATLAB are written such that when the argument (input) is an array, the operation that is defined by the function is executed on each element of the array. (One can think about the operation as element-by-element application of the function.) The result (output) from such an operation is an array in which each element is calculated by entering the corresponding element of the argument (input) array into the function. For example, if a vector with seven elements is substituted in the function cos (x), the result is a vector with seven elements in which each element is the cosine of the corresponding element in x. This is shown below:

```
>> x=[0:pi/6:pi]

x =

    0    0.5236    1.0472    1.5708    2.0944    2.6180    3.1416

>>y=cos(x)

y =
  1.0000    0.8660    0.5000    0.0000   -0.5000   -0.8660   -1.0000
>>
```

An example in which the argument variable is a matrix is:

```
>> d=[1 4 9; 16 25 36; 49 64 81]          Creating a 3 × 3 array.

d =

     1     4     9
    16    25    36
    49    64    81

>> h=sqrt(d)

h =
     1     2     3               h is a 3 × 3 array in which each
     4     5     6               element is the square-root of the
     7     8     9               corresponding element in array d.
```

The feature of MATLAB, in which arrays can be used as arguments in functions, is called vectorization.

3.6 BUILT-IN FUNCTIONS FOR ANALYZING ARRAYS

MATLAB has many built-in functions for analyzing arrays. Table 3-1 lists some of these functions.

Table 3-1: Built-in array functions

Function	Description	Example

`mean(A)`	If A is a vector, returns the mean value of the elements of the vector.	`>> A=[5 9 2 4];`
		`>> mean(A)`
		`ans =`
		    `5`

`C=max(A)`	If A is a vector, C is the largest element in A. If A is a matrix, C is a row vector containing the largest element of each column of A.	`>> A=[5 9 2 4 11 6 11 1];`
		`>> C=max(A)`
		`C =`
		    `11`

`[d,n]=max(A)`	If A is a vector, d is the largest element in A, n is the position of the element (the first if several have the max value).	`>> [d,n]=max(A)`
		`d =`
		    `11`
		`n =`
		    `5`

`min(A)`	The same as `max(A)`, but for the smallest element.	`>> A=[5 9 2 4];`
		`>> min(A)`
		`ans =`
		    `2`

`[d,n]=min(A)`	The same as `[d,n]= max(A)`, but for the smallest element.	

`sum(A)`	If A is a vector, returns the sum of the elements of the vector.	`>> A=[5 9 2 4];`
		`>> sum(A)`
		`ans =`
		    `20`

`sort(A)`	If A is a vector, arranges the elements of the vector in ascending order.	`>> A=[5 9 2 4];`
		`>> sort(A)`
		`ans =`
		    `2` `4` `5` `9`

`median(A)`	If A is a vector, returns the median value of the elements of the vector.	`>> A=[5 9 2 4];`
		`>> median(A)`
		`ans =`
		    `4.5000`

Table 3-1: Built-in array functions (Continued)

Function	Description	Example

std(A)	If A is a vector, returns the standard deviation of the elements of the vector.	`>> A=[5 9 2 4];`
		`>> std(A)`
		`ans =`
		`    2.9439`

det(A)	Returns the determinant of a square matrix A.	`>> A=[2 4; 3 5];`
		`>> det(A)`
		`ans =`
		`    -2`

dot(a,b)	Calculates the scalar (dot) product of two vectors a and b. The vectors can each be row or column vectors.	`>> a=[1 2 3];`
		`>> b=[3 4 5];`
		`>> dot(a,b)`
		`ans =`
		`    26`

cross(a,b)	Calculates the cross product of two vectors a and b, (a×b). The two vectors must have 3 elements.	`>> a=[1 3 2];`
		`>> b=[2 4 1];`
		`>> cross(a,b)`
		`ans =`
		`    -5    3    -2`

inv(A)	Returns the inverse of a square matrix A.	`>> A=[2 -2 1; 3 2 -1; 2 -3 2];`
		`>> inv(A)`
		`ans =`
		`    0.2000   0.2000        0`
		`   -1.6000   0.4000   1.0000`
		`   -2.6000   0.4000   2.0000`

3.7 GENERATION OF RANDOM NUMBERS

Simulations of many physical processes and engineering applications frequently requires using a number (or a set of numbers) that has a random value. MATLAB has two commands rand and randn that can be used to assign random numbers to variables.

The rand command:

The rand command generates uniformly distributed numbers with values between 0 and 1. The command can be used to assign these numbers to a scalar, a vector, or a matrix, as shown in Table 3-2.

Table 3-2: The `rand` command

Command	Description	Example

rand	Generates a single random number between 0 and 1.	`>> rand`
		`ans =`
		`    0.2311`

rand(1,n)	Generates an n element row vector of random numbers between 0 and 1.	`>> a=rand(1,4)`
		`a =`
		`  0.6068  0.4860  0.8913  0.7621`

rand(n)	Generates an n × n matrix with random numbers between 0 and 1.	`>> b=rand(3)`
		`b =`
		`   0.4565   0.4447   0.9218`
		`   0.0185   0.6154   0.7382`
		`   0.8214   0.7919   0.1763`

rand(m,n)	Generates an m × n matrix with random numbers between 0 and 1.	`>> c=rand(2,4)`
		`c =`
		`  0.4057  0.9169  0.8936  0.3529`
		`  0.9355  0.4103  0.0579  0.8132`

randperm(n)	Generates a row vector with n elements that are random permutation of integers 1 through n.	`>> randperm(8)`
		`ans =`
		`   8   2   7   4   3   6   5   1`

Sometimes there is a need to have random numbers that are distributed in an interval other than (0,1), or to have numbers that are only integers. This can be done by mathematical operations with the `rand` function. Random numbers that are distributed in a range (a,b) can be obtained by multiplying rand by $(b - a)$ and adding the product to a:

$$(b - a)*\text{rand} + a$$

For example, a vector of 10 elements with random values between –5 and 10 can be created by ($a = -5$, $b = 10$):

```
>> v=15*rand(1,10)-5
v =
   -1.8640    0.6973    6.7499    5.2127    1.9164    3.5174
  6.9132   -4.1123    4.0430   -4.2460
```

Random numbers that are all integers can be generated by using one of the rounding functions. For example, a 2 × 15 matrix with random integers with values that range from 1 to 100 can be created by:

```
>> A=round(99*rand(2,15)+1)
A =
    24    6   64   85   18   45   32   40   13   46   93   17   25   97   87
    25    9   20   18   99   35   37   60    5   87   27   87   65   67    2
```

The `randn` command:

The `randn` command generates normally distributed numbers with mean 0 and standard deviation of 1. The command can be used to generate a single number, a vector, or a matrix in the same way as the `rand` command. For example, a 3×4 matrix is created by:

```
>> d=randn(3,4)
d =
    -0.4326     0.2877     1.1892     0.1746
    -1.6656    -1.1465    -0.0376    -0.1867
     0.1253     1.1909     0.3273     0.7258
```

The mean and standard deviation of the numbers can be changed by mathematical operations to have any values. This is done by multiplying the number generated by the `randn` function by the desired standard deviation, and adding the desired mean. For example, a vector of 15 numbers (integers) with a mean of 50 and standard deviation of 5 is generated by:

```
>> v=round(5*randn(1,15)+50)
v =
    53   51   45   39   50   45   53   53   58   53   47   52   45   50   50
```

In the example above integers are obtained by using the `round` function.

3.8 EXAMPLES OF MATLAB APPLICATIONS

Sample Problem 3-2: Equivalent force system (addition of vectors)

Three forces are applied to a bracket as shown. Determine the total (equivalent) force applied to the bracket.

Solution

A force is a vector (physical quantity that has a magnitude and direction). In a Cartesian coordinate system a two-dimensional vector **F** can be written as:

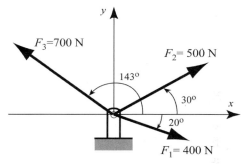

$$\mathbf{F} = F_x\mathbf{i} + F_y\mathbf{j} = F\cos\theta\mathbf{i} + F\sin\theta\mathbf{j} = F(\cos\theta\mathbf{i} + \sin\theta\mathbf{j})$$

where F is the magnitude of the force, and θ is its angle relative to the x axis, F_x and F_y are the components of **F** in the directions of the x and y axis, respectively, and **i** and **j** are unit vectors in these directions. If F_x and F_y are known, then F and θ can be determined by:

$$F = \sqrt{F_x^2 + F_y^2} \quad \text{and} \quad \tan\theta = \frac{F_y}{F_x}$$

The total (equivalent) force applied on the bracket is obtained by adding the forces that are acting on the bracket. The MATLAB solution below follows three steps:

- Write each force as a vector with two elements, where the first element is the x component of the vector and the second element is the y component.

- Determine the vector form of the equivalent force by adding the vectors.

- Determine the magnitude and direction of the equivalent force.

The problem is solved in the following script file.

```
% Sample Problem 3-2 solution (script file)
clear
F1M=400; F2M=500; F3M=700;
```
Define variables with the magnitude of each vector.

```
Th1=-20; Th2=30; Th3=143;
```
Define variables with the angle of each vector.

```
F1=F1M*[cosd(Th1) sind(Th1)]
F2=F2M*[cosd(Th2) sind(Th2)]
F3=F3M*[cosd(Th3) sind(Th3)]
```
Define the three vectors.

```
Ftot=F1+F2+F3
```
Calculate the total force vector.

```
FtotM=sqrt(Ftot(1)^2+Ftot(2)^2)
```
Calculate the magnitude of the total force vector.

```
Th=atand(Ftot(2)/Ftot(1))
```
Calculate the angle of the total force vector.

When the program is executed, the following is displayed in the Command Window:

```
F1 =
   375.8770  -136.8081
```
The components of F1

```
F2 =
   433.0127   250.0000
```
The components of F2

```
F3 =
  -559.0449   421.2705
```
The components of F3

```
Ftot =
   249.8449   534.4625
FtotM =
   589.9768
Th =
   64.9453
```

The components of the total force.

The magnitude of the total force.

The direction of the total force.

The equivalent force has a magnitude of 589.98 N, and is directed 64.95° (ccw) relative to the x axis. In vector notation the force is: $\mathbf{F} = 249.84\mathbf{i} + 534.46\mathbf{j}$ N.

Sample Problem 3-3:   Friction experiment (element-by-element calculations)

The coefficient of friction, μ, can be determined in an experiment by measuring the force F required to move a mass m. When F is measured and m is known, the coefficient of friction can be calculated by:

$$\mu = F/(mg) \quad (g = 9.81 \text{ m/s}^2).$$

friction

Results from measuring F in six tests are given in the table below. Determine the coefficient of friction in each test, and the average from all tests.

Test #	1	2	3	4	5	6
Mass m (kg)	2	4	5	10	20	50
Force F (N)	12.5	23.5	30	61	117	294

Solution

A solution using MATLAB commands in the Command Window is shown below.

```
>> m=[2 4 5 10 20 50];
>> F=[12.5 23.5 30 61 117 294];
>> mu=F./(m*9.81)

mu =
   0.6371    0.5989    0.6116    0.6218    0.5963    0.5994

>> mu_ave=mean(mu)

mu_ave =
   0.6109
```

Enter the values of m in a vector.

Enter the values of F in a vector.

A value for mu is calculated for each test, using element-by-element calculations.

The average of the elements in the vector mu is determined by using the function mean.

Sample Problem 3-4:   Electrical resistive network analysis (solving a system of linear equations)

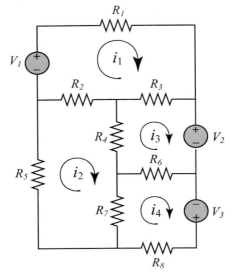

The electrical circuit shown to the right consists of resistors and voltage sources. Determine the current in each resistor using the mesh current method which is based on Kirchhoff's voltage law.

$V_1 = 20$ V, $V_2 = 12$ V, $V_3 = 40$ V
$R_1 = 18\,\Omega$, $R_2 = 10\,\Omega$, $R_3 = 16\,\Omega$
$R_4 = 6\,\Omega$, $R_5 = 15\,\Omega$, $R_6 = 8\,\Omega$
$R_7 = 12\,\Omega$, $R_8 = 14\,\Omega$

Solution

Kirchhoff's voltage law states that the sum of the voltage around a closed circuit is zero. In the mesh current method a current is first assigned for each mesh (i_1, i_2, i_3, i_4 in the figure). Then, Kirchhoff's voltage second law is applied for each mesh. This results in a system of linear equations for the currents (in our case four equations). The solution gives the values of the mesh currents. The current in a resistor that belongs to two meshes is the sum of the currents in the corresponding meshes. It is convenient to assume that all the currents are in the same direction (clockwise in our case). In the equation for each mesh, the voltage source is positive if the current flows to the – pole, and the voltage of a resistor is negative for current in the direction of the mesh current.

The equations for the four meshes in the current problem are:

$$V_1 - R_1 i_1 - R_3(i_1 - i_3) - R_2(i_1 - i_2) = 0$$
$$-R_5 i_2 - R_2(i_2 - i_1) - R_4(i_2 - i_3) - R_7(i_2 - i_4) = 0$$
$$-V_2 - R_6(i_3 - i_4) - R_4(i_3 - i_2) - R_3(i_3 - i_1) = 0$$
$$V_3 - R_8 i_4 - R_7(i_4 - i_2) - R_6(i_4 - i_3) = 0$$

The four equations can be rewritten in matrix form $[A][x] = [B]$:

$$
\begin{bmatrix}
-(R_1 + R_2 + R_3) & R_2 & R_3 & 0 \\
R_2 & -(R_2 + R_4 + R_5 + R_7) & R_4 & R_7 \\
R_3 & R_4 & -(R_3 + R_4 + R_6) & R_6 \\
0 & R_7 & R_6 & -(R_6 + R_7 + R_8)
\end{bmatrix}
\begin{bmatrix}
i_1 \\ i_2 \\ i_3 \\ i_4
\end{bmatrix}
=
\begin{bmatrix}
-V_1 \\ 0 \\ V_2 \\ -V_3
\end{bmatrix}
$$

The problem is solved in the following program written in a script file:

```
V1=20; V2=12; V3=40;
R1=18; R2=10; R3=16; R4=6;          Define variables with the
R5=15; R6=8; R7=12; R8=14;          values of the V's and R's.
A=[-(R1+R2+R3) R2 R3 0
R2 -(R2+R4+R5+R7) R4 R7             Create the matrix A.
R3 R4 -(R3+R4+R6) R6
0 R7 R6 -(R6+R7+R8)]
>> B=[-V1; 0; V2; -V3]              Create the vector B.
>> I=A\B                           Solve for the currents by using left division.
```

When the script file is executed, the following is displayed in the Command Window:

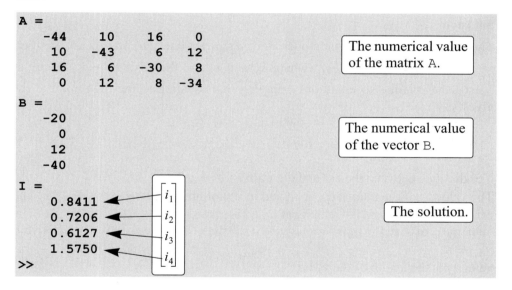

```
A =
    -44     10     16      0
     10    -43      6     12        The numerical value
     16      6    -30      8        of the matrix A.
      0     12      8    -34

B =
    -20
      0                             The numerical value
     12                             of the vector B.
    -40

I =
    0.8411   ← i₁
    0.7206   ← i₂                   The solution.
    0.6127   ← i₃
    1.5750   ← i₄
>>
```

The last column vector gives the current in each mesh. The currents in the resistors R_1, R_5, and R_8 are $i_1 = 0.8411$ A, $i_2 = 0.7206$ A, and $i_4 = 1.5750$ A, respectively. The other resistors belong to two meshes and their current is the sum of the currents in the meshes.

The current in resistor R_2 is $i_1 - i_2 = 0.1205$ A.

The current in resistor R_3 is $i_1 - i_3 = 0.2284$ A.

The current in resistor R_4 is $i_2 - i_3 = 0.1079$ A.

The current in resistor R_6 is $i_4 - i_3 = 0.9623$ A.

The current in resistor R_7 is $i_4 - i_2 = 0.8544$ A.

Sample Problem 3-5:   Motion of two particles

A train and a car are approaching a road crossing. At
$t = 0$ the train is 400 ft. south of the crossing traveling
north at a constant speed of 54 mi/h. At the same
time the car is 200 ft. west of the crossing traveling
east at a speed of 28 mi/h and accelerating at 4 ft/s².
Determine the positions of the train and the car, the
distance between them, and the speed of the train relative to the car every second for the next 10 seconds.

To show the results, create an 11 × 6 matrix in
which each row has the time in the first column and
the train position, car position, distance between the
train and the car, car speed, and the speed of the train
relative to the car, in the next five columns, respectively.

Solution

The position of an object that moves along a straight line at a constant acceleration
is given by $s = s_o + v_o t + \frac{1}{2}at^2$ where s_o and v_o are the position and velocity at
$t = 0$, and a is the acceleration. Applying this equation to the train and the car
gives:

$$y = -400 + v_{otrain}t \quad \text{(train)}$$
$$x = -200 + v_{ocar}t + \frac{1}{2}a_{car}t^2 \quad \text{(car)}$$

The distance between the car and the train is: $d = \sqrt{x^2 + y^2}$.
The velocity of the train is constant and in vector notation is: $\mathbf{v}_{train} = v_{otrain}\mathbf{j}$. The
car is accelerating and its velocity at time t is given by: $\mathbf{v}_{car} = (v_{ocar} + a_{car}t)\mathbf{i}$. The
velocity of the train relative to the car, $\mathbf{v}_{t/c}$ is given by:
$\mathbf{v}_{t/c} = \mathbf{v}_{train} - \mathbf{v}_{car} = -(v_{ocar} + a_{car}t)\mathbf{i} + v_{otrain}\mathbf{j}$. The magnitude (speed) of this
velocity is the length of the vector.

The problem is solved in the following program written in a script file. First a
vector t with 11 elements for the time from 0 to 10 s is created, then the positions
of the train and the car, the distance between them, and the speed of the train relative to the car at each time element are calculated.

```
v0train=54*5280/3600;  v0car=28*5280/3600;  acar=4;
```
Create variables for the initial velocities (in ft/s) and the acceleration.
```
t=0:10;
```
Create the vector t.
```
y=-400+v0train*t;
x=-200+v0car*t+0.5*acar*t.^2;
```
Calculate the train and car positions.
```
d=sqrt(x.^2+y.^2);
```
Calculate the distance between the train and car.
```
vcar=v0car+acar*t;
```
Calculate the car's velocity.

```
speed_trainRcar=sqrt(vcar.^2+v0train^2);
```
Calculate the speed of the train relative to the car.
```
table=[t' y' x' d' vcar' speed_trainRcar']
```
Create a table (see note below).

Note: In the commands above, `table` is the name of the variable that is a matrix containing the data to be displayed.

When the script file is executed, the following is displayed in the Command Window:

```
table =
         0 -400.0000 -200.0000   447.2136   41.0667    89.2139
    1.0000 -320.8000 -156.9333   357.1284   45.0667    91.1243
    2.0000 -241.6000 -109.8667   265.4077   49.0667    93.1675
    3.0000 -162.4000  -58.8000   172.7171   53.0667    95.3347
    4.0000  -83.2000   -3.7333    83.2837   57.0667    97.6178
    5.0000   -4.0000   55.3333    55.4777   61.0667   100.0089
    6.0000   75.2000  118.4000   140.2626   65.0667   102.5003
    7.0000  154.4000  185.4667   241.3239   69.0667   105.0849
    8.0000  233.6000  256.5333   346.9558   73.0667   107.7561
    9.0000  312.8000  331.6000   455.8535   77.0667   110.5075
   10.0000  392.0000  410.6667   567.7245   81.0667   113.3333
```

Time (s)	Train position (ft)	Car position (ft)	Car-train distance (ft)	Car speed (ft/s)	Train speed relative to the car; (ft/s)

In this problem the results (numbers) are displayed by MATLAB without any text. Instructions on how to add text to output generated by MATLAB are presented in Chapter 4.

3.9 PROBLEMS

Note: Additional problems for practicing mathematical operations with arrays are provided at the end of Chapter 4.

1. For the function $y = \dfrac{(2x^2 - 5x + 4)^3}{x^2}$, calculate the value of y for the following values of x: $-2, -1, 0, 1, 2, 3, 4, 5$ using element-by-element operations.

2. For the function $y = 5\sqrt{t} - \dfrac{(t+2)^2}{0.5(t+1)} + 8$, calculate the value of y for the following values of t: $0, 1, 2, 3, 4, 5, 6, 7, 8$ using element-by-element operations.

3. A ball that is dropped on the floor bounces back up many times, reaching a lower height after each bounce. When the ball impacts the floor its rebound velocity is 0.85 times the impact velocity. The velocity v that a ball hits the floor after being dropped from a height h is given by $v = \sqrt{2gh}$, where $g = 9.81$ m/s^2 is the acceleration of the Earth. The maximum height h_{max} that a ball reaches is given by $h_{max} = \dfrac{v^2}{2g}$, where v is the

upper velocity after impact. Consider a ball that is dropped from a height of 2 m. Determine the height the ball reaches after the first 8 bounces. (Calculate the velocity of the ball when it hit the floor for the first time. Derive a formula for h_{max} as a function of the bounce number. Then create a vector $n = 1, 2, ..., 8$ and use the formula (use element-by element operations) to calculate a vector with the values of h_{max} for each n.)

4. If a basketball is dropped down from a helicopter, its velocity as a function of time $v(t)$ can be modeled by the equation:

$$v(t) = \sqrt{\frac{2mg}{\rho A C_d}}\left(1 - e^{-\sqrt{\frac{\rho g C_d A}{2m}}\, t}\right)$$

where $g = 9.81$ m/s^2 is the gravitation of the Earth, $C_d = 0.5$ is the drag coefficient, $\rho = 1.2$ kg/m^3 is the density of air, $m = 0.624$ kg is the mass of the basketball, and $A = \pi r^2$ is the projected area of the ball ($r = 0.117$ m is the radius). Determine the velocity of the basketball for $t = 0, 1, 2, 3, 4, 5, 6, 7, 8, 9,$ and 10 s. Note that initially the velocity increases rapidly, but then due to the resistance of the air, the velocity increases more gradually. Eventually the velocity approaches a limit that is called the terminal velocity.

5. The length $|\mathbf{u}|$ (magnitude) of a vector $\mathbf{u} = x\mathbf{i} + y\mathbf{j} + z\mathbf{k}$ is given by $|\mathbf{u}| = \sqrt{x^2 + y^2 + z^2}$. Given the vector $\mathbf{u} = 14\mathbf{i} + 25\mathbf{j} - 10\mathbf{k}$, determine its length two ways:
   a) Define the vector in MATLAB, and then write a mathematical expression that uses the components of the vector.
   b) Define the vector in MATLAB, then use element-by element operation to create a new vector with elements that are the square of the original vector. Then use MATLAB built-in functions sum and sqrt to calculate the length. All of these can be written in one command.

6. The position as a function of time $(x(t), y(t))$ of a projectile fired with a speed of v_0 at an angle θ is given by:

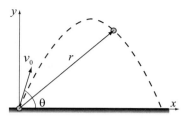

$$x(t) = v_0 \cos\theta \cdot t \qquad y(t) = v_0 \sin\theta \cdot t - \frac{1}{2}gt^2$$

where $g = 9.81 \text{ m/s}^2$ is the gravitation of the Earth. The distance r to the projectile at time t can be calculated by $r(t) = \sqrt{x(t)^2 + y(t)^2}$. Consider the case where $v_0 = 100 \text{ m/s}$ and $\theta = 79°$. Determine the distance r to the projectile for $t = 0, 2, 4, ..., 20$ s.

7. Two vectors are given:

   $\mathbf{u} = 4\mathbf{i} + 9\mathbf{j} - 5\mathbf{k}$ and $\mathbf{v} = -3\mathbf{i} + 6\mathbf{j} - 7\mathbf{k}$

   Use MATLAB to calculate the dot product $\mathbf{u} \cdot \mathbf{v}$ of the vectors in two ways:

   a) Define $\mathbf{u}$ as a row vector and $\mathbf{v}$ as a column vector, and then use matrix multiplication.
   b) Use MATLAB built-in function `dot`.

8. Define x and y as the vectors $x = 2, 4, 6, 8, 10$ and $y = 3, 6, 9, 12, 15$. Then use them in the following expression to calculate z using element-by-element calculations.

$$z = \left(\frac{y}{x}\right)^2 + (x+y)^{\left(\frac{y-x}{x}\right)}$$

9. Define h and k as scalars, $h = 0.7$, and $k = 8.85$, and x, y and z as the vectors $x = [1, 2, 3, 4, 5]$, $y = [2.1, 2.0, 1.9, 1.8, 1.7]$, and $z = [2.0, 2.5, 3.0, 3.5, 4.0]$. Then use these variables to calculate G using element-by-element calculations for the vectors.

$$G = \frac{hx + ky}{(x+y)^h} + \frac{e^{\left(\frac{hy}{z}\right)}}{z^{(y/x)}}$$

10. Show that $\lim\limits_{x \to 0} \dfrac{e^x - 1}{x} = 1$

   Do this by first creating a vector x that has the elements: 1 0.5 0.1 0.01 0.001 0.00001 and 0.0000001. Then, create a new vector y in which each element is determined from the elements of x by $\dfrac{e^x - 1}{x}$. Compare the elements of y with the value 1 (use format long to display the numbers.

11. Use MATLAB to show that the sum of the infinite series $4 \sum_{n=0}^{\infty} \frac{(-1)^n}{2n+1}$ con-

    verges to π. Do it by computing the sum for:
    a) $n = 100$
    b) $n = 10{,}000$
    c) $n = 1{,}000{,}000$

    For each part create a vector n in which the first element is 0, the increment is 1 and the last term is 100, 10,000, or 1,000,000. Then, use element-by-element calculation to create a vector in which the elements are $\frac{(-1)^n}{2n+1}$. Finally, use MATLAB built-in function sum to add the terms of the series (and multiply the sum by 4). Compare the values obtained in parts a, b, and c with the value of π. (Don't forget to type semicolons at the end of commands that otherwise will display large vectors.)

12. Use MATLAB to show that the sum of the infinite series $\sum_{n=0}^{\infty} \frac{1}{(2n+1)(2n+2)}$

    converges to ln 2. Do this by computing the sum for:
    a) $n = 50$
    b) $n = 500$
    c) $n = 5{,}000$

    For each part create a vector n in which the first element is 0, the increment is 1 and the last term is 50, 500, or 5,000. Then, use element-by-element calculation to create a vector in which the elements are $\frac{1}{(2n+1)(2n+2)}$. Finally, use the function sum to add the terms of the series. Compare the values obtained in parts a, b, and c to ln 2.

13. Fisheries commonly estimate the growth of a fish population using the von Bertalanffy growth law:

    $$L = L_{max}(1 - e^{-K(t+\tau)})$$

    where L_{max} is the maximum length, K is a rate constant, and τ is a time constant. These constants vary with species of fish. Assume $L_{max} = 50\,cm$, and $\tau = 0.5$ years, calculate the length of a fish at 2 years of age for $K = 0.25$, 0.5, and 0.75 years^{-1}.

14. Create the following three matrices:

$$A = \begin{bmatrix} 5 & 2 & 4 \\ 1 & 7 & -3 \\ 6 & -10 & 0 \end{bmatrix} \qquad B = \begin{bmatrix} 11 & 5 & -3 \\ 0 & -12 & 4 \\ 2 & 6 & 1 \end{bmatrix} \qquad C = \begin{bmatrix} 7 & 14 & 1 \\ 10 & 3 & -2 \\ 8 & -5 & 9 \end{bmatrix}$$

  a) Calculate $A + B$ and $B + A$ to show that addition of matrices is commutative.

  b) Calculate $A + (B + C)$ and $(A + B) + C$ to show that addition of matrices is associative.

  c) Calculate $5(A + C)$ and $5A + 5C$ to show that, when matrices are multiplied by a scalar, the multiplication is distributive.

  d) Calculate $A*(B + C)$ and $A*B + A*C$ to show that matrix multiplication is distributive.

15. Use the matrices A, B, and C from the previous problem to answer the following:

  a) Does $A*B = B*A$?

  b) Does $A*(B*C) = (A*B)*C$?

  c) Does $(A*B)^t = B^t*A^t$? (^{t} means transpose)

  d) Does $(A + B)^t = A^t + B^t$?

16. Two projectiles, A and B, are shot at the same instant from the same spot. Projectile A is shot at a speed of 680 m/s at an angle of $65°$ and projectile B is shot at a speed of 780 m/s at an angle of $42°$. Determine which projectile will hit the ground first. Then, take the flying time t_f of that projectile and divide it into ten increments by creating a vector t with 11 equally spaced elements (the first element is 0, the last is t_f). Calculate the distance between the two projectiles at the eleven times in vector t.

17. The mechanical power output P in a contracting muscle is given by:

$$P = Tv = \frac{kvT_0\left(1 - \dfrac{v}{v_{max}}\right)}{k + \dfrac{v}{v_{max}}}$$

where T is the muscle tension, v is the shortening velocity (max of v_{max}), T_0 is the isometric tension (i.e. tension at zero velocity) and k is a nondimensional constant that ranges between 0.15 and 0.25 for most muscles. The equation can be written in nondimensional form:

$$p = \frac{ku(1-u)}{k+u}$$

where $p = (Tv)/(T_0 v_{max})$, and $u = v/v_{max}$. A figure with $k = 0.25$ is shown on the right.

a) Create a vector u ranging from 0 to 1 with increments of 0.05.

b) Using $k = 0.25$, calculate the value of p for each value of u.

c) Using MATLAB built-in function `max`, find the maximum value of p.

d) Repeat the first three steps with increments of 0.01 and calculate the per-cent relative error defined by: $E = \left|\dfrac{p_{max_{0.01}} - p_{max_{0.05}}}{p_{max_{0.05}}}\right| \times 100$.

18. Solve the following system of five linear equations:

$$1.5x - 2y + z + 3u + 0.5w = 7.5$$
$$3x + y - z + 4u - 3w = 16$$
$$2x + 6y - 3z - u + 3w = 78$$
$$5x + 2y + 4z - 2u + 6w = 71$$
$$-3x + 3y + 2z + 5u + 4w = 54$$

19. The electrical circuit shown consists of resistors and voltage sources. Determine the current in each resistor, using the mesh current method that is based on Kirchhoff's second voltage law (see Sample Problem 3-4).

$V_1 = 38$ V, $V_2 = 20$ V, $V_3 = 24$ V
$R_1 = 15\,\Omega$, $R_2 = 18\,\Omega$, $R_3 = 10\,\Omega$
$R_4 = 9\,\Omega$, $R_5 = 5\,\Omega$, $R_6 = 14\,\Omega$
$R_7 = 8\,\Omega$, $R_8 = 13\,\Omega$

Chapter 4

Using Script Files and Managing Data

A script file (see Section 1.8) is a list of MATLAB commands, called a program, that is saved in a file. When the script file is executed (run), MATLAB executes the commands. Section 1.8 describes how to create, save, and run a simple script file in which the commands are executed in the order that they are listed, and in which all the variables are defined within the script file. The present chapter gives more details of how to input data to a script file, how data is stored in MATLAB, various ways to display and save data that is created in script files, and how to exchange data between MATLAB and other applications. (How to write more advanced programs where commands are not necessarily executed in a simple order is covered in Chapter 7.)

In general, variables can be defined (created) in several ways. As shown in Chapter 2, variables can be defined implicitly by assigning values to a variable name. Variables can also be assigned values by the output of a function. In addition, variables can be defined with data that is imported from files outside MAT-LAB. Once defined (either in the Command Window or when a script file is executed) the variables are stored in MATLAB's Workspace.

Variables that reside in the Workspace can be displayed in different ways, saved, or exported to applications outside MATLAB. Similarly, data from files outside MATLAB can be imported to the Workspace and then used in MATLAB.

Section 4.1 explains how MATLAB stores data in the Workspace and how the user can see the data that is stored. Section 4.2 shows how variables that are used in script files can be defined in the Command Window and/or in Script files. Section 4.3 shows how to output data that is generated when script files are executed. Section 4.4 explains how the variables in the Workspace can be saved and then retrieved, and Section 4.5 shows how to import and export data from and to applications outside MATLAB.

4.1 THE MATLAB WORKSPACE AND THE WORKSPACE WINDOW

The MATLAB workspace contains of the set of variables (named arrays) that are defined and stored during a MATLAB session. It includes variables that have been defined in the Command Window and variables defined when script files are executed. This means that the Command Window and script files share the same memory zone within the computer. This implies that once a variable is in the workspace, it is recognized and can be used, and can be reassigned new values, in both the Command Window and script files. As will be explained in Chapter 6 (Section 6.3), there is another type of files in MATLAB, called function files, where variables can also be defined. These variables, however, are normally not shared with other parts of the program since they use a separate workspace.

Recall from Chapter 1 that the who command displays a list of the variables currently in the workspace. The whos command displays a list of the variables currently in the workspace and information about their size, bytes, and class. An example is shown below.

```
>> 'Variables in memory'                          Typing a string.
ans =
Variables in memory                      The string is assigned to ans.
>> a = 7;
>> E = 3;                                 Creating the variables a,
>> d = [5,   a+E,   4,   E^2]             E, d, and g.
d =
      5    10    4    9
>> g = [a, a^2,  13;  a*E,  1,  a^E]
g =
      7    49    13
     21     1    343
>> who                             The who command displays the vari-
                                   ables currently in the workspace.
Your variables are:
E    a    ans d    g
>> whos                                       The whos com-
  Name       Size           Bytes  Class      mand displays
                                              the variables
    E        1x1               8   double array   currently in the
    a        1x1               8   double array   workspace, and
    ans      1x19             38   char array     information
    d        1x4             32   double array    about their size.
    g        2x3             48   double array

Grand total is 31 elements using 134 bytes
>>
```

The variables currently in memory can also be viewed in the Workspace Window. If not open, this window can be opened by selecting **Workspace** in the **Desktop** menu. Figure 4-1 shows the Workspace Window that corresponds to the variables that were defined above. The variables that are displayed in the Work-

Figure 4-1: The Workspace Window.

space Window can also be edited (changed). Double-clicking on a variable opens the Array Editor Window, where the content of the variable is displayed in a table. For example, Figure 4-2 shows the Array Editor Window that opens when the variable g in Figure 4-1 is double-clicked.

Figure 4-2: The Array Editor Window.

The elements in the Array Editor Window can be edited. The variables in the Workspace Window can be deleted by selecting them, and then pressing the **delete** key on the keyboard, or selecting **delete** from the **edit** menu. This has the same effect as entering the command `clear variable_name` in the Command Window.

4.2 INPUT TO A SCRIPT FILE

When a script file is executed the variables that are used in the calculations within the file must have assigned values. In other words, the variables must be in the workspace. The assignment of a value to a variable can be done in three ways, depending on where and how the variable is defined.

1. The variable is defined and assigned value in the script file.

In this case the assignment of value to the variable is part of the script file. If the user wants to run the file with a different variable value, the file must be edited and the assignment of the variable changed. Then, after the file is saved, it can be executed again.

The following is an example of such a case. The script file (saved as Chapter4Example2) calculates the average points scored in three games.

```
% This script file calculates the average points scored in three games.
% The assignment of the values of the points is part of the script file.
game1=75;
game2=93;                              The variables are assigned
game3=68;                              values within the script file.
ave_points=(game1+game2+game3)/3
```

The Command Window when this file is executed looks like:

```
>> Chapter4Example2
                    The script file is executed by typing the name of the file.
ave_points =
   78.6667          The variable ave_points with its value
>>                  is displayed in the Command Window.
```

2. The variable is defined and assigned value in the Command Window.

In this case the assignment of a value to the variable is done in the Command Window. (Recall that the variable is recognized in the script file.) If the user wants to run the script file with a different value for the variable, the new value is assigned in the Command Window and the file is executed again.

For the previous example in which the script file has a program that calculates the average of points scored in three games, the script file (saved as Chapter4Example3) is:

```
% This script file calculates the average points scored in three games.
% The assignment of the values of the points to the variables
% game1, game2, and game3 is done in the Command Window.

ave_points=(game1+game2+game3)/3
```

The Command Window for running this file is:

```
>> game1 = 67;
>> game2 = 90;           The variables are assigned values in
>> game3 = 81;           the Command Window.
```

```
>> Chapter4Example3
```
The script file is executed.

```
ave_points =
    79.3333
```
The output from the script file is displayed in the Command Window.

```
>> game1 = 87;
>> game2 = 70;
>> game3 = 50;
```
New values are assigned to the variables.

```
>> Chapter4Example3
```
The script file is executed again.

```
ave_points =
      69
>>
```
The output from the script file is displayed in the Command Window.

3. The variable is defined in the script file, but a specific value is entered in the Command Window when the script file is executed.

In this case the variable is defined in the script file and when the file is executed, the user is prompted to assign a value to the variable in the Command Window. This is done by using the `input` command to create the variable.

The form of the `input` command is:

```
variable_name = input('string with a message that
                is displayed in the Command Window')
```

When the `input` command is executed as the script file runs, the string is displayed in the Command Window. The string is a message prompting the user to enter a value that is assigned to the variable. The user types the value and presses the **Enter** key. This assigns the value to the variable. As with any variable, the variable and its assigned value will be displayed in the Command Window unless a semicolon is typed at the very end of the `input` command. A script file that uses the `input` command to enter the points scored in each game to the program that calculates the average of the scores is shown below.

```
% This script file calculates the average of points scored in three games.
% The point from each game are assigned to the variables by
% using the input command.
game1=input('Enter the points scored in the first game ');
game2=input('Enter the points scored in the second game ');
game3=input('Enter the points scored in the third game ');
ave_points=(game1+game2+game3)/3
```

The following shows the Command Window when this script file (saved as

Chapter4Example4) is executed.

```
>> Chapter4Example4
Enter the points scored in the first game    67
Enter the points scored in the second game   91
Enter the points scored in the third game    70


ave_points =
    76
>>
```

> The computer displays the message. Then the value of the score is typed by the user and the **Enter** key is pressed.

In this example scalars are assigned to the variables. In general, however, vectors and arrays can also be assigned. This is done by typing the array in the same way that it is usually assigned to a variable (left bracket, then typing row by row, and a right bracket).

The `input` command can also be used to assign a string to a variable. This can be done in one of two ways. One way is to use the command in the same form as shown above, and when the prompt message appears the string is typed in between two single quotes in the same way that a string is assigned to a variable without the `input` command. The second way is to use an option in the `input` command that defines the characters that are entered as a string. The form of the command is:

> variable_name = input ('prompt message', 's')

where the 's' inside the command defines the characters that will be entered as a string. In this case when the prompt message appears, the text is typed in without the single quotes, but it is assigned to the variable as a string. An example where the `input` command is used with this option is included in Sample Problem 7-4.

4.3 OUTPUT COMMANDS

As discussed before, MATLAB automatically generates a display when some commands are executed. For example, when a variable is assigned a value, or the name of a previously assigned variable is typed and the **Enter** key is pressed, MATLAB displays the variable and its value. This type of output is not displayed if a semicolon is typed at the end of the command. In addition to this automatic display, MATLAB has several commands that can be used to generate displays. The displays can be messages that provide information, numerical data, and plots. Two commands that are frequently used to generate output are the `disp` and `fprintf`. The `disp` command displays the output on the screen, while the `fprintf` command can be used to display the output on the screen, or to save the output to a file. The commands can be used in the Command Window, in a script file, and, as will be shown later, in a function file. When these commands are used in a script file, the display output that they generate is displayed in the Command Window.

4.3.1 The disp Command

The disp command is used to display the elements of a variable without displaying the name of the variable, and to display text. The format of the disp command is:

> disp(name of a variable) or disp('text as string')

- Every time the disp command is executed, the display it generates appears in a new line. One example is:

```
>> abc = [5  9  1;  7  2  4];
```
A 2 × 3 array is assigned to variable abc.

```
>> disp(abc)
```
The disp command is used to display the abc array.

```
       5       9       1
       7       2       4
```
The array is displayed without its name.

```
>> disp('The problem has no solution.')

The problem has no solution.
>>
```
The disp command is used to display a message.

The next example shows the use of the disp command in the script file that calculates the average points scored in three games.

```
% This script file calculates the average points scored in three games.
% The points from each game are assigned to the variables by
% using the input command.
% The disp command is used to display the output.

game1=input('Enter the points scored in the first game    ');
game2=input('Enter the points scored in the second game   ');
game3=input('Enter the points scored in the third game    ');
ave_points=(game1+game2+game3)/3;
disp(' ')
disp('The average of points scored in a game is:')
disp(' ')
disp(ave_points)
```

Display empty line.
Display text.
Display empty line.
Display the value of the variable ave_points.

When this file (saved as Chapter4Example5) is executed, the display in the

Command Window is:

```
>> Chapter4Example5
Enter the points scored in the first game    89
Enter the points scored in the second game   60
Enter the points scored in the third game    82
```
An empty line is displayed.

```
The average of points scored in a game is:
```
The text line is displayed.

An empty line is displayed.

```
    77
```
The value of the variable `ave_points` is displayed.

- Only one variable can be displayed in a disp command. If elements of two variables need to be displayed together, a new variable (that contains the elements to be displayed) must first be defined and then displayed.

In many situations it is nice to display output (numbers) in a table. This can be done by first defining a variable that is an array with the numbers and then using the disp command to display the array. Headings to the columns can also be created with the disp command. Since in the disp command the user cannot control the format (the width of the columns and the distance between the columns) of the display of the array, the position of the headings has to be adjusted to the columns by adding spaces. As an example, the script file below shows how to display the population data from Chapter 2 in a table.

```
yr=[1984 1986 1988 1990 1992 1994 1996];
```
The population data is entered in two row vectors.
```
pop=[127 130 136 145 158 178 211];
```

```
tableYP(:,1)=yr';
```
yr is entered as the first column in the array `tableYP`.
```
tableYP(:,2)=pop';
```
pop is entered as the second column in the array `tableYP`.
```
disp('           YEAR        POPULATION')
```
Display heading (first line).
```
disp('                       (MILLIONS)')
```
Display heading (second line).
```
disp(' ')
```
Display an empty line.
```
disp(tableYP)
```
Display the array `tableYP`.

When this script file (saved as PopTable) is executed the display in the Command Window is:

```
>> PopTable
        YEAR        POPULATION
                    (MILLIONS)
```
Headings are displayed.

An empty line is displayed.

```
        1984          127
        1986          130
```

1988	136	The `tableYP` array is displayed.
1990	145	
1992	158	
1994	178	
1996	211	

Another example of displaying a table is shown in Sample Problem 4-3. Tables can also be created and displayed with the `fprintf` command, which is explained in the next section.

4.3.2 The `fprintf` Command

The `fprintf` command can be used to display output (text and data) on the screen or to save it to a file. With this command (unlike with the `disp` command) the output can be formatted. For example, text and numerical values of variables can be intermixed and displayed in the same line. In addition, the format of the numbers can be controlled.

With many available options, the `fprintf` command can be long and complicated. To avoid confusion, the command is presented gradually. First, it is shown how to use the command to display text messages, then how to mix numerical data and text, followed by how to format the display of numbers, and lastly how to save the output to a file.

Using the `fprintf` command to display text:

To display text, the `fprintf` command has the form:

```
fprintf('text typed in as a string')
```

For example:

```
fprintf('The problem, as entered, has no solution. Please check the
input data.')
```

If this line is part of a script file, when the line is executed, the following is displayed in the Command Window:

```
The problem, as entered, has no solution. Please check the input data.
```

With the `fprintf` command it is possible to start a new line in the middle of the string. This is done by inserting `\n` before the character that will start the new line. For example, inserting `\n` after the first sentence in the previous example gives:

```
fprintf('The problem, as entered, has no solution.\nPlease
check the input data.')
```

When this line executes, the display in the Command Window is:

```
The problem, as entered, has no solution.
Please check the input data.
```

The \n is called an escape character. It is used to control the display. Other escape characters that can be inserted within the string are:

    \b   Backspace.
    \t   Horizontal tab.

When a program has more than one fprintf command, the display that they generate is continuous (the fprintf command does not automatically start a new line). This is true even if there are other commands between the fprintf commands. An example is the following script file:

```
fprintf('The problem, as entered, has no solution. Please check the
input data.')
x = 6; d = 19 + 5*x;
fprintf('Try to run the program later.')
y = d + x;
fprintf('Use different input values.')
```

When this file is executed the display in the Command Window is:

```
The problem, as entered, has no solution. Please check the
input data.Try to run the program later.Use different input
values.
```

To start a new line with the fprintf command, a \n must be typed at the start of the string.

Using the fprintf command to display a mix of text and numerical data:

To display a mix of text and a number (value of a variable), the fprintf command has the form:

The % sign marks the spot where the number is inserted within the text.

Formatting elements (define the format of the number).

The name of the variable whose value is displayed.

The formatting elements are:

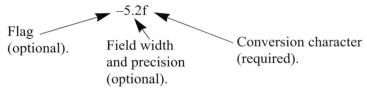

The flag, which is optional can be one of the following three:

Character used for flag	Description
– (minus sign)	Left justifies the number within the field.
+ (plus sign)	Prints a sign character (+ or –) in front of the number.
0 (zero)	Adds zeros if the number is shorter than the field.

The field width and precision (5.2 in the previous example) are optional. The first number (5 in the example) is the field width that specifies the minimum number of digits in the display. If the number to be displayed is shorter than the field width, spaces or zeros are added in front of the number. The precision is the second number (2 in the previous example) specifies the number of digits to be displayed to the right of the decimal point.

The last element in the formatting elements, which is required, is the conversion character which specifies the notation in which the number is displayed. Some of the common notations are:

e	Exponential notation using lower case e (e.g. 1.709098e+001).
E	Exponential notation using upper case E (e.g. 1.709098E+001).
f	Fixed point notation (e.g. 17.090980).
g	The shorter of e or f notations.
G	The shorter of E or f notations.
i	Integer.

Information about additional notation is available in the help menu of MATLAB. As an example, the `fprintf` command with a mix of text and a number is used in the script file that calculates the average points scored in three games.

```
% This script file calculates the average points scored in three games.
% The values are assigned to the variables by using the input command.
% The fprintf command is used to display the output.
game(1) = input('Enter the points scored in the first game   ');
game(2) = input('Enter the points scored in the second game   ');
game(3) = input('Enter the points scored in the third game   ');
ave_points = mean(game);
```

Notice that, besides using the `fprintf` command, this file differs from the ones shown earlier in the chapter in that the scores are stored in the first three elements of a vector named `game`, and the average of the scores is calculated by using the `mean` function. The Command Window where the script file above (saved as Chapter4Example6) was run is shown below.

```
>> Chapter4Example6
Enter the points scored in the first game    75
Enter the points scored in the second game   60
Enter the points scored in the third game    81
An average of 72.000000 points was scored in the three games.
>>
```

> The display generated by the `fprintf` command combines text and a number (value of a variable).

With the `fprintf` command it is possible to insert more than one number (values of a variable) within the text. This is done by typing %g (or % followed by any formatting elements) at the places in the text where the numbers are to be inserted. Then, after the string argument of the command (following the comma), the names of the variables are typed in the order that they are inserted in the text. In general the command looks like:

```
fprintf('..text...%g...%g...%f...',variable1,variable2,variable3)
```

An example is shown in the following script file:

```
% This program calculates the distance a projectile flies,
% given its initial velocity and the angle at which it is shot.
% the fprintf command is used to display a mix of text and numbers.

v=1584;   % Initial velocity (km/h)
theta=30;   % Angle (degrees)
vms=v*1000/3600;
t=vms*sind(30)/9.81;
```

> Changing velocity units to m/s.

> Calculating the time to highest point.

```
d=vms*cosd(30)*2*t/1000;
```
Calculating max distance.
```
fprintf('A projectile shot at %3.2f degrees with a velocity
of %4.2f km/h will travel a distance of %g km.\n',theta,v,d)
```

When this script file (saved as Chapter4Example7) is executed, the display in the Command Window is:

```
>> Chapter4Example7
A projectile shot at 30.00 degrees with a velocity of
1584.00 km/h will travel a distance of 17.091 km.
>>
```

Additional Remarks About the `fprintf` Command:

- To have a single quotation mark in the displayed text, type two single quotation marks in the string inside the command.

- The `fprintf` command is vectorized. This means that when a variable that is a vector or a matrix is included in the command, the command repeats itself until all the elements are displayed. If the variable is a matrix the data is used column by column.

    For example, the script file below creates a 2×5 matrix T in which the first row are the numbers 1 through 5, and the second row are the corresponding square roots.

```
x=1:5;
```
Create a vector x.
```
y=sqrt(x);
```
Create a vector y.
```
T=[x; y]
```
Create 2×5 matrix T, first row is x, second row is y.
```
fprintf('If the number is: %i, its square root is: %f\n',T)
```
The `fprintf` command displays two numbers from T in every line.

When this script file is executed the display in the Command Window is:

```
T =
    1.0000    2.0000    3.0000    4.0000    5.0000
    1.0000    1.4142    1.7321    2.0000    2.2361
```
The 2×5 matrix T.
```
If the number is: 1, its square root is: 1.000000
If the number is: 2, its square root is: 1.414214
If the number is: 3, its square root is: 1.732051
If the number is: 4, its square root is: 2.000000
If the number is: 5, its square root is: 2.236068
```
The `fprintf` command repeats 5 times, using the numbers from the matrix T column after column.

Using the `fprintf` **command to save output to a file:**

In addition to displaying output in the Command Window, the `fprintf` command can be used for writing the output to a file when it is necessary to save the output. The data that is saved can subsequently be displayed or used in MATLAB and in other applications.

Writing output to a file requires three steps:

a) Opening a file using the `fopen` command.
b) Writing the output to the open file using the `fprintf` command.
c) Closing the file using the `fclose` command.

Step *a*:

Before data can be written to a file, the file must be opened. This is done with the `fopen` command, which creates a new file or opens an existing file. The `fopen` command has the form:

$$ fid = fopen('file\_name','permission') $$

`fid` is a variable called the file identifier. A scalar value is assigned to `fid` when `fopen` is executed. The file name is written (including its extension) within single quotes as a string. The permission is a code (also written as a string) that tells how the file is opened. Some of the more common permission codes are:

'r'   Open file for reading (default).
'w'   Open file for writing. If the file already exists, its content is deleted. If the file does not exists, a new file is created.
'a'   Same as 'w', except that if the file exists the written data is appended to the end of the file.

If a permission code is not included in the command, the file opens with the default code 'r'. Additional permission codes are described in the help menu.

Step *b*:

Once the file is open, the `fprintf` command can be used to write output to the file. The `fprintf` command is used in exactly the same way as it is used to display output in the Command Window, except that the variable `fid` is inserted inside the command. The `fprintf` command then has the form:

```
fprintf(fid, 'text %-5.2f additional text',vari
                                    able_name)
```

`fid` is added to the `fprintf` command.

Step *c*:

When writing the data to the file is complete, the file is closed using the `fclose` command. The `fclose` command has the form:

$$\boxed{\texttt{fclose(fid)}}$$

Additional notes on using the `fprintf` **command for saving output to a file:**

- The created file is saved in the current directory.

- It is possible to use the `fprintf` command to write to several different files. This is done by first opening the files, assigning a different `fid` to each (e.g. `fid1`, `fid2`, `fid3`, etc.), and then using the `fid` of a specific file in the `fprintf` command to write to that file.

An example of using `fprintf` commands for saving output to two files is shown in the following script file. The program in the file generates two unit conversion tables. One table converts velocity units from miles per hour to kilometers per hour, and the other table converts force units from pounds to Newtons. Each conversion table is saved to a different text file (extension .txt).

```
% Script file in which fprintf is used to write output to files.
% Two conversion tables are created and saved to two different files.
% One converts mi/h to km/h, the other converts lb to N.
clear all
Vmph=10:10:100;                  Creating a vector of velocities in mph.
Vkmh=Vmph.*1.609;                      Converting mph to km/h.
TBL1=[Vmph; Vkmh];               Creating a table (matrix) with two rows.
Flb=200:200:2000;                    Creating a vector of forces in lb.
FN=Flb.*4.448;                         Converting lb to N.
TBL2=[Flb; FN];                  Creating a table (matrix) with two rows.
fid1=fopen('VmphtoVkm.txt','w')  Open a txt file named VmphtoVkm.
fid2=fopen('FlbtoFN.txt','w');     Open a txt file named FlbtoFN.
fprintf(fid1,'Velocity Conversion Table\n \n');
                                 Writing a title and an empty line to the file fid1.
fprintf(fid1,'     mi/h          km/h    \n');
                                 Writing two columns heading to the file fid1.
fprintf(fid1,'   %8.2f        %8.2f\n',TBL1);
                           Writing the data from the variable TBL1 to the file fid1.
```

```
fprintf(fid2,'Force Conversion Table\n \n');
fprintf(fid2,'    Pounds        Newtons    \n');
fprintf(fid2,'   %8.2f        %8.2f\n',TBL2);
fclose(fid1);
fclose(fid2);
```

> Writing the force conversion table (data in variable `TBL2`) to the file fid2.

> Closing the files fid1 and fid2.

When the script file above is executed two new txt files, named Vmph-toVkm and FlbtoFN, are created and saved in the current directory. These files can be opened with any application that can read txt files. Figures 4-3 and 4-4 show how the two files appear when they are opened with Microsoft Word.

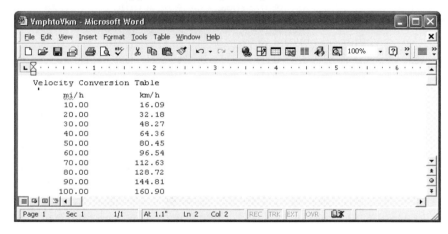

Figure 4-3: The Vmphtokm.txt file opened in Word.

Figure 4-4: The FlbtoFN.txt file opened in Word.

4.4 THE **save** AND **load** COMMANDS

The `save` and `load` commands are most useful for saving and retrieving data for use in MATLAB. The `save` command can be used for saving the variables that are currently in the workspace, and the `load` command is used for retrieving variables that have been previously saved, back to the workspace. The workspace can be saved when MATLAB is used in one type of platform (e.g. PC), and retrieved for use in MATLAB in another platform (e.g. Mac). The `save` and `load` commands can also be used for exchanging data with applications outside MATLAB. Additional commands that can be used for this purpose are presented in Section 4.5.

4.4.1 The `save` Command

The `save` command is used for saving the variables (all or some) that are stored in the workspace. The two simplest forms of the `save` command are:

```
save file_name
```
   or
```
save('file_name')
```

When the either one of these commands is executed, all the variables currently in the workspace are saved in a file named `file_name.mat` that is created in the current directory. In `mat` files, which are written in a binary format, each variable preserves its name, type, size, and value. These files cannot be read by other applications. The `save` command can also be used for saving only some of the variables that are in the workspace. For example, to save two variables named `var1` and `var2` the command is:

```
save file_name var1 var2
```
   or

```
save('file_name','var1','var2')
```

The `save` command can also be used for saving in ASCII format, which can be read by applications outside MATLAB. Saving in ASCII format is done by adding the argument `-ascii` in the command (for example: `save file_name -ascii`). In the ASCII format the variable's name, type, and size are not preserved. The data is saved as characters separated by spaces but without the variable names. For example, the following shows how two variables (a 1×4 vector and a 2×3 matrix) are defined in the Command Window, and then are saved in ASCII format to a file named `DatSaveAsci`:

```
>> V=[3 16 -4 7.3];              Create a 1 × 4 vector V.
>> A=[6 -2.1 15.5; -6.1 8 11];   Create a 1 × 4 matrix A.
>> save -ascii DatSavAsci        Save variables to a file named DatSavAs.
```

Once saved, the file can be opened with any application that can read ASCII files. For example, Figure 4-5 shows the data when the file is opened with Notepad.

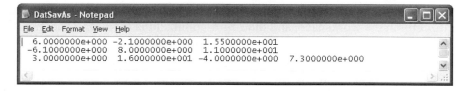

Figure 4-5: Data saved in ASCII format.

Note that the file does not include the name of the variables. The numerical values of the variables (first A and then V) are just listed.

4.4.2 *The* load *Command*

The load command can be used for retrieving variables that were saved with the save command back to the workspace, and for importing data that was created with other applications and saved in ASCII format or in text (.txt) files. Variables that were saved with the save command in .mat files can be retrieved with the commands:

> load file_name   or   load('file_name')

When the commands are executed, all the variables in the file (with the name, type, size and values as were saved) are added (loaded back) to the workspace. If the workspace has already a variable with the same name as a variable that is retrieved with the load command, then the variable that is retrieved replaces the existing variable. The load command can also be used for retrieving only some of the variables that are in the saved .mat file. For example, to retrieve two variables named var1 and var2, the command is:

> load file_name var1 var2   or
>
>   load('file_name','var1','var2')

The load command can also be used to import data that is saved in ASCII or text (.txt) to the workspace. This is possible, however, only if the data in the file is in a form of a variable in MATLAB. Thus, the file can have one number (scalar), a row or a column of numbers (vectors), or rows with the same number of numbers in each (matrix). For example, the data shown in Figure 4-5 cannot be loaded with the load command (even though it was saved in ASCII format with the save command) because the number of elements is not the same in all the rows. (Recall that this file was created by saving two different variables.)

When data is loaded from an ASCII or a text file into the workspace it has to be assigned to a variable name. Data in an ASCII format can be loaded with the following two forms of the `load` command:

`load file_name`   or   `VarName=load('file_name')`

If the data is in a text file the extension `.txt` has to be added to the file name. The form of the `load` command is then:

`load file_name.txt` or `VarName=load('file_name.txt')`

In the first form of the commands the data is assigned to a variable that has the name of the file. In the second form the data is assigned to a variable named `Var-Name`.

For example, the data shown in Figure 4-6 (a 3×2 matrix) is typed in Notepad, and then saved as `DataFromText.txt`.

Figure 4-6: Data saved as txt file.

Next, two forms of the `load` command are used to import the data in the text file to the Workspace of MATLAB. In the first command the data is assigned to a variable named `DfT`. In the second command the data is automatically assigned to a variable named `DataFromText`, which is the name of the text file where the data was saved.

```
>> DfT=load('DataFromText.txt')
DfT =
    56.0000    -4.2000
     3.0000     7.5000
    -1.6000   198.0000
>> load DataFromText.txt
>> DataFromText
DataFromText =
    56.0000    -4.2000
     3.0000     7.5000
    -1.6000   198.0000
```

Load the file `DataFrom-Text` and assign the loaded data to the variable `Dft`.

Use the load command with the file `DataFromText`.

The data is assigned to a variable named `DataFromText`.

Importing data to (or exporting from) other applications can also be done with other MATLAB commands that are presented in the next section.

4.5 IMPORTING AND EXPORTING DATA

MATLAB is often used for analyzing data that was recorded in experiments or generated by other computer programs. This can be done by first importing the data into MATLAB. Similarly, data that is produced by MATLAB sometimes needs to be transferred to other computer applications. There are various types of data (numerical, text, audio, graphics, and images). This section only describes how to import and export numerical data which is probably the most common type of data that need to be transferred by new users of MATLAB. For other types of data transfer, the user should look in the Help Window under File I/O.

Importing data can be done either by using commands, or by using the Import Wizard. Commands are useful when the format of the data being imported is known. MATLAB has several commands that can be used for importing various types of data. Importing commands can also be included in a script file such that the data is imported when the script is executed. The Import Wizard is useful when the format of the data (or the command that is applicable for importing the data) is not known. The Import Wizard determines the format of the data and automatically imports it.

4.5.1 Commands for Importing and Exporting Data

This section describes – in detail – how to transfer data into and out of Excel spreadsheets. Microsoft Excel is commonly used for storing data, and Excel is compatible with many data recording devices and computer applications. Many people are also capable of importing and exporting various data formats into and from Excel. MATLAB has also commands for transferring data directly to and from formats like csv and ASCII, and to the spreadsheet program Lotus 123. Details of these and many other commands can be found in the Help Window under File I/O

Importing and exporting data into and from Excel:

Importing data from Excel is done with the `xlsread` command. When the command is executed, the data from the spreadsheet is assigned as an array to a variable. The simplest form of the `xlsread` command is:

```
variable_name=xlsread('filename')
```

- `'filename'` (typed as a string) is the name of the Excel file. The directory of the Excel file must either be the current directory, or listed in the search path.

- If the Excel file has more than one sheet, the data will be imported from the first sheet.

When an Excel file has several sheets, the `xlsread` command can be used to import data from a specified sheet. The form of the command is then:

$$\boxed{\texttt{variable\_name = xlsread('filename', 'sheet\_name')}}$$

- The name of the sheet is typed as a string.

Another option is to import only a portion of the data that is in the spreadsheet. This is done by typing an additional argument in the command:

$$\boxed{\texttt{variable\_name = xlsread('filename', 'sheet\_name', 'range')}}$$

- The 'range' (typed as a string) is a rectangular region of the spreadsheet defined by the addresses (in Excel notation) of the cells at opposite corners of the region. For example, 'C2:E5' is a 4 by 3 region of rows 2, 3, 4, and 5 and columns *C*, *D*, and *E*.

Exporting data from MATLAB to an Excel spreadsheet is done by using the `xlswrite` command. The simplest form of the command is:

$$\boxed{\texttt{xlswrite('filename', variable\_name)}}$$

- 'filename' (typed as a string) is the name of the Excel file to which the data is exported. The file must be in the current directory. If the file does not exist, a new Excel file with the specified name will be created.

- `variable_name` is the name of the variable in MATLAB with the assigned data that is being exported.

- The arguments 'sheet_name' and 'range' can be added to the `xlswrite` command to export to a specified sheet and to a specified range of cells, respectively.

As an example, the data from the Excel spreadsheet shown in Figure 4-7 is imported into MATLAB by using the `xlsread` command.

Figure 4-7: Excel spreadsheet with data.

The spreadsheet is saved in a file named TestData1 in a disk in drive A. After the Current Directory is changed to drive A, the data is imported into MATLAB by assigning it to the variable DATA:

```
>> DATA = xlsread('TestData1')
DATA =
   11.0000    2.0000   34.0000   14.0000   -6.0000         0    8.0000
   15.0000    6.0000  -20.0000    8.0000    0.5600   33.0000    5.0000
    0.9000   10.0000    3.0000   12.0000  -25.0000   -0.1000    4.0000
   55.0000    9.0000    1.0000   -0.5550   17.0000    6.0000  -30.0000
```

4.5.2 Using the Import Wizard

Using the Import Wizard is probably the easiest way to import data into MAT-LAB since the user does not have to know, or to specify, the format of the data. The Import Wizard is activated by selecting **Import Data** in the **File** menu of the Command Window. (It can also be started by typing the command uiimport.) The Import Wizard starts by displaying a file selection box that shows all the data files recognized by the Wizard. The user then selects the file which contains the data to be imported, and clicks **Open**. The Import Wizard opens the file and displays a portion of the data in a preview box of the wizard such that the user can verify that the data is the correct one. The Import Wizard tries to process the data, and if the wizard is successful, it displays the variables it has created with a portion of the data. The user clicks **next** and the wizard shows the Column Separator that was used. If the variable has the correct data, the user can proceed with the wizard (click **next**); otherwise the user can choose a different Column Separator. In the next window the wizard shows the name and size of the variable to be created in MATLAB. (When the data is all numerical, the variable in MATLAB has the same name as the file from which the data was imported.) When the wizard ends (click **finish**), the data is imported to MATLAB.

As an example, the Import Wizard is used to import numerical ASCII data saved in a txt file. The data saved with the file name TestData2 is shown in Figure 4-8.

Figure 4-8: Numerical ASCII data.

The display of the Import Wizard during the import process of the TestData2 file is shown in Figures 4-9 through 4-11. The third figure shows that the name of the variable in MATLAB is `TestData2`, and its size is 3 by 5.

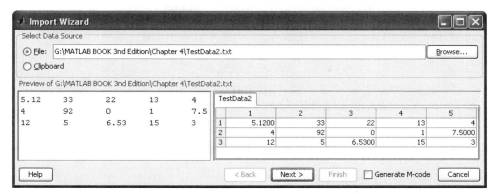

Figure 4-9: Import Wizard, first display.

Figure 4-10: Import Wizard, second display.

Figure 4-11: Import Wizard, third display.

In the Command Window of MATLAB, the imported data can be displayed by typing the name of the variable.

```
>> TestData2

TestData2 =
    5.1200   33.0000   22.0000   13.0000    4.0000
    4.0000   92.0000         0    1.0000    7.5000
   12.0000    5.0000    6.5300   15.0000    3.0000
```

4.6 EXAMPLES OF MATLAB APPLICATIONS

Sample Problem 4-1:   Height and surface area of a silo

A cylindrical silo with radius r has a spherical cap roof with radius R. The height of the cylindrical portion is H. Write a program in a script file that determines the height H for given values of r, R, and the volume V. In addition, the program also calculates the surface area of the silo.

Use the program to calculate the height and surface area of a silo with r = 30 ft., R = 45 ft., and a volume of 120,000 ft^3. Assign values for r, R, and V in the Command Window.

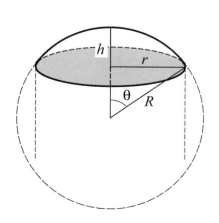

Solution

The total volume of the silo is obtained by adding the volume of the cylindrical part and the volume of the spherical cap. The volume of the cylinder is given by:

$$V_{cyl} = \pi r^2 H$$

and the volume of the spherical cap is given by:

$$V_{cap} = \frac{1}{3}\pi h^2(3R - h)$$

where $h = R - R\cos\theta = R(1 - \cos\theta)$,

and θ is calculated from $\sin\theta = \dfrac{r}{R}$.

Using the equations above, the height, H, of the cylindrical part can be expressed by:

$$H = \frac{V - V_{cap}}{\pi r^2}$$

The surface area of the silo is obtained by adding the surface areas of the cylindrical part and the spherical cap.

$$S = S_{cyl} + S_{cap} = 2\pi rH + 2\pi Rh$$

A program in a script file that solves the problem is presented below:

```
theta=asin(r/R);                      Calculating θ.
h=R*(1-cos(theta));                   Calculating h.
Vcap=pi*h^2*(3*R-h)/3;           Calculating the volume of the cap.
H=(V-Vcap)/(pi*r^2);                  Calculating H.
S=2*pi*(r*H + R*h);          Calculating the surface area S.
fprintf('The height H is: %f ft.',H)
fprintf('\nThe surface area of the silo is: %f square ft.',S)
```

The Command Window where the script file, named silo, was executed is:

```
>> r=30; R=45; V=200000;      Assigning values to r, R, and V.
>> silo                        Running the script file named silo.
The height H is: 64.727400 ft.
The surface area of the silo is: 15440.777753 square ft.
```

Sample Problem 4-2: Centroid of a composite area

Write a program in a script file that calculates the coordinates of the centroid of a composite area. (A composite area can easily be divided into sections whose centroids are known.) The user needs to divide the area to sections and know the coordinates of the centroid (two numbers) and the area of each section (one number). When the script file is executed, it asks the user to enter the three numbers as a row in a matrix. The user enters as many rows as there are sections. A section that represents a hole is taken to have a negative area. For output, the program displays the coordinates of the centroid of the composite area. Use the program to calculate the centroid of the area shown in the figure.

Dimensions in mm.

Solution

The area is divided into six sections as shown in the following figure. The total

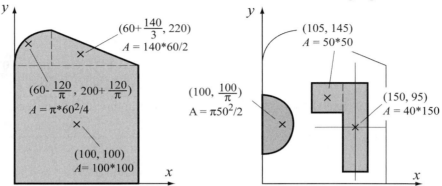

Units: coordinates mm, area mm².

area is calculated by adding the three sections on the left and subtracting the three sections on the right. The location and coordinates of the centroid of each section are marked in the figure, as well as the area of each section.

The coordinates $\bar{X}$ and $\bar{Y}$ of the centroid of the total area are given by:

$\bar{X} = \dfrac{\Sigma A\bar{x}}{\Sigma A}$ and $\bar{Y} = \dfrac{\Sigma A\bar{y}}{\Sigma A}$, where $\bar{x}$, $\bar{y}$, and A are the coordinates of the centroid and area of each section, respectively.

A script file with a program for calculating the coordinates of the centroid of a composite area is:

```
% The program calculates the coordinates of the centroid
% of a composite area.
clear C xs ys As
C=input('Enter a matrix in which each row has three ele-
ments.\nIn each row enter the x and y coordinates of the
centroid and the area of a section.\n');
xs=C(:,1)';
ys=C(:,2)';
As=C(:,3)';
A=sum(As);
x=sum(As.*xs)/A;
y=sum(As.*ys)/A;
fprintf('The coordinates of the centroid are: ( %f, %f )',x,y)
```

xs=C(:,1)'; → Creating a row vector for the *x* coordinate of each section (first column of C).

ys=C(:,2)'; → Creating a row vector for the *y* coordinate of each section (second column of C).

As=C(:,3)'; → Creating a row vector for the area of each section (third column of C).

A=sum(As); → Calculating the total area.

x=sum(As.*xs)/A; → Calculating the coordinates of the centroid of the composite area.
y=sum(As.*ys)/A;

The script file was saved with the name Centroid. The following shows the Command Window where the script file was executed.

```
>> Centroid
Enter a matrix in which each row has three elements.

In each row enter the x and y coordinates of the centroid
and the area of a section.
[100   100   200*200
60-120/pi   200+120/pi   pi*60^2/4
60+140/3   220   140*60/2                  Entering the data for matrix C.
100   100/pi   -pi*50^2/2                  Each row has three elements: the
150   95   -40*150                         x, y, and A of a section.
105   145   -50*50]

The   coordinates   of   the   centroid   are:   (   85.387547   ,
131.211809 )
```

Sample Problem 4-3: Voltage divider

When several resistors are connected in an electrical circuit in series, the voltage across each of them is given by the voltage divider rule:

$$v_n = \frac{R_n}{R_{eq}} v_s$$

where v_n and R_n are the voltage across resistor n and its resistance, respectively, $R_{eq} = \Sigma R_n$ is the equivalent resistance, and v_s is the source voltage. The power dissipated in each resistor is given by:

$$P_n = \frac{R_n}{R_{eq}^2} v_s^2$$

The figure below shows, for example, a circuit with seven resistors connected in series.

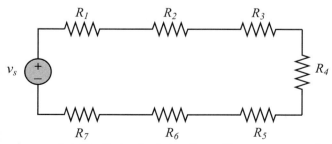

Write a program in a script file that calculates the voltage across each resistor, and the power dissipated in each resistor, in a circuit that has resistors connected in series. When the script file is executed it requests the user to first enter the source voltage and then to enter the resistances of the resistors in a vector. The program displays a table with the resistances listed in the first column, the voltage across the resistor in the second, and the power dissipated in the resistor in the third column. Following the table, the program displays the current in the circuit, and the total power.

Execute the file and enter the following data for v_s and the R's.

$$v_s = 24\text{V}, \quad R_1 = 20\,\Omega, \quad R_2 = 14\,\Omega, \quad R_3 = 12\,\Omega, \quad R_4 = 18\,\Omega, \quad R_5 = 8\,\Omega,$$
$$R_6 = 15\,\Omega, \quad R_7 = 10\,\Omega.$$

Solution

A script file that solves the problem is shown below.

```
% The program calculates the voltage across each resistor
% in a circuit that has resistors connected in series.
vs=input('Please enter the source voltage ');
Rn=input('Enter the values of the resistors as elements in a
row vector\n');
Req=sum(Rn);                    Calculate the equivalent resistance.
vn=Rn*vs/Req;                   Apply the voltage divider rule.
Pn=Rn*vs^2/Req^2;               Calculate the power in each resistor.
i = vs/Req;                     Calculate the current in the circuit.
Ptotal = vs*i;                  Calculate the total power in the circuit.
Table = [Rn', vn', Pn'];        Create a variable table with the
disp(' ')                       vectors Rn, vn, and Pn as columns.
disp(' Resistance Voltage    Power')   Display headings for
disp('    (Ohms)    (Volts)   (Watts)')  the columns.
disp(' ')                       Display an empty line.
disp(Table)                     Display the variable Table.
disp(' ')
fprintf('The current in the circuit is %f Amps.',i)
fprintf('\nThe total power dissipated in the circuit is %f
Watts.',Ptotal)
```

The Command Window where the script file was executed is:

```
>> VoltageDivider                       Name of the script file.
Please enter the source voltage 24      Voltage entered by the user.
Enter the value of the resistors as elements in a row vector

[20  14  12  18  8  15  10]             Resistor values entered as a vector.
```

Resistance (Ohms)	Voltage (Volts)	Power (Watts)
20.0000	4.9485	1.2244
14.0000	3.4639	0.8571
12.0000	2.9691	0.7346
18.0000	4.4536	1.1019
8.0000	1.9794	0.4897
15.0000	3.7113	0.9183
10.0000	2.4742	0.6122

The current in the circuit is 0.247423 Amps.
The total power dissipated in the circuit is 5.938144 Watts.

4.7 PROBLEMS

Solve the following problems by first writing a program in a script file and then executing the program.

1. A paper cup shaped as a frustum of cone with $R_2 = 1.25R_1$ is designed to have a volume of 250 cm³. Determine R_1, R_2, and the surface area, S, of the paper for cups with heights h, of 5, 6, 7, 8, 9, and 10 cm. The volume of the cup, V, and the surface area of the paper are given by:

$$V = \frac{1}{3}\pi h(R_1^2 + R_2^2 + R_1 R_2)$$

$$S = \pi(R_1 + R_2)\sqrt{(R_2 - R_1)^2 + h^2} + \pi R_1^2$$

2. In a movie theater the angle θ at which a viewer sees the picture on the screen depends on the distance x of the viewer from the screen. For a movie theater with the dimensions shown in the figure, determine the angle θ (in degrees) for viewers sitting at distances of 30, 45, 60, 75, and 90 ft. from the screen.

3. A student has a summer job as a life-
    guard at the beach. After spotting a
    swimmer in trouble he tries to deduce
    the path by which he can reach the
    swimmer in the shortest time. The
    path of shortest distance (path A) is
    obviously not best since it maximizes
    the time spent swimming (he can run
    faster than he can swim). Path B mini-

    mizes the time spent swimming but is probably not best since it is the longest
    (reasonable) path. Clearly the optimal path is somewhere in between paths A
    and B.

    Consider an intermediate path C and determine the time required to reach
    the swimmer in terms of the running speed v_{run} = 3 m/s, swimming speed
    v_{swim} = 1 m/s, the distances L = 48 m, d_s = 30 m, and d_w = 42 m, and the
    lateral distance y at which the lifeguard enters the water. Create a vector y that
    ranges between path A and path B (y = 20, 21, 22, ..., 48 m) and compute a
    time t for each y. Use MATLAB built-in function min to find the minimum
    time t_{min} and the entry point y for which it occurs. Determine the angles that
    correspond to the calculated value of y and investigate whether your result sat-
    isfies Snell's law of refraction:

    $$\frac{\sin\phi}{\sin\alpha} = \frac{v_{run}}{v_{swim}}$$

4. Radioactive decay of radioactive materials can be modeled by the equation
    $A = A_0 e^{kt}$, where A is the population at time t, A_0 is the amount at $t = 0$, and
    k is the decay constant ($k \le 0$). Technetium-99 is a radioisotope that is used in
    imaging of the brain. Its half-life time is 6 hours. Calculate the relative
    amount of Technetium-99 (A/A_0) in a patient body for 24 hours after receiv-
    ing a dose. After determining the value of k, define a vector t = 0, 2, 4, ..., 24
    and calculate the corresponding values of A/A_0.

5. Write a script file that determines the balance in a saving account at the end of
    every year for the first 10 years. The account has an initial investment of
    $1,000 and interest rate of 6.5% that compounds annually. Display the infor-
    mation in a table.

    For an initial investment of A, and interest rate of r, the balance B after n
    years is given by:

    $$B = A\left(1 + \frac{r}{100}\right)^n$$

6. The velocity, v, and the distance, d, as a function of time, of a car that accelerates from rest at constant acceleration, a, are given by:

$$v(t) = at \quad \text{and} \quad d(t) = \frac{1}{2}at^2$$

Determine v and d as every second for the first 10 seconds for a car with acceleration of $a = 1.55$ m/s^2. Display the results in a three-column table in which the first column is time (s), the second distance (m), and the third is velocity (m/s).

7. The variation of vapor pressure p (in units of mm Hg) of benzene with temperature in the range of $0 \leq T \leq 42$ °C can be modeled with the equation (Handbook of Chemistry and Physics, CRC Press):

$$\log_{10} p = b - \frac{-0.05223a}{T}$$

where $a = 34,172$ and $b = 7.9622$ are material constants and T is absolute temperature (K). Write a program in a script file that calculates the pressure for various temperatures. The program creates a vector of temperatures from $T = 0$°C to $T = 42$°C with increments of 2 degrees, and displays a two-columns table of p and T, where the first column are temperatures in °C, and the second column are the corresponding pressures in mm Hg.

8. The temperature dependence of the heat capacity C_p of many gases can be described in terms of a cubic equation:

$$C_p = a + bT + cT^2 + dT^3$$

The following table gives the coefficients of the cubic equation for four gases. C_p is in Joules/(g mol)(°C) and T is in °C.

gas	a	b	c	d
SO_2	38.91	3.904×10^{-2}	-3.105×10^{-5}	8.606×10^{-9}
SO_3	48.50	9.188×10^{-2}	-8.540×10^{-5}	32.40×10^{-9}
O_2	29.10	1.158×10^{-2}	-0.6076×10^{-5}	1.311×10^{-9}
N_2	29.00	0.2199×10^{-2}	-0.5723×10^{-5}	-2.871×10^{-9}

Calculate the heat capacity for each gas at temperatures ranging between 200 and 400°C at 20°C increments. To present the results, create an 11×5 matrix where the first column is the temperature, and the second through fifth columns are the heat capacities of SO_2, SO_3, O_2, and N_2, respectively.

9. The heat capacity of an ideal mixture of four gases $C_{p_{mixture}}$ can be expressed in terms of the heat capacity of the components by the mixture equation:

$$C_{p_{mixture}} = x_1 C_{p1} + x_2 C_{p2} + x_3 C_{p3} + x_4 C_{p4}$$

where x_1, x_2, x_3, and x_4 are the fraction of the components, and C_{p1}, C_{p2}, C_{p3}, and C_{p4} are the corresponding heat capacities. A mixture of unknown quantities of the four gases SO_2, SO_3, O_2, and N_2 is given. To determine the fraction of the components, the following values of the heat capacity of the mixture were measured at three temperatures:

Temperature °C	25	150	300
$C_{p_{mixture}}$ Joules/(g mol)(°C)	39.82	44.72	49.10

Use the equation and data in the previous problem for determining the heat capacity of the four components at the three temperatures. Then, use the mixture equation to write three equations for the mixture at the three temperatures. The fourth equation is: $x_1 + x_2 + x_3 + x_4 = 1$. Determine x_1, x_2, x_3, and x_4 by solving the linear system of equations.

10. When several resistors are connected in an electrical circuit in parallel, the current through each of them is given by $i_n = \dfrac{v_s}{R_n}$ where i_n and R_n are the current through resistor n and its resistance, respectively, and v_s is the source voltage. The equivalent resistance, R_{eq}, can be determined from the equation:

$$\frac{1}{R_{eq}} = \frac{1}{R_1} + \frac{1}{R_2} + \dots + \frac{1}{R_n}$$

The source current is given by: $i_s = v_s/R_{eq}$, and the power, P_n, dissipated in each resistor is given by: $P_n = v_s i_n$.

Write a program in a script file that calculates the current through each resistor and the power dissipated in each in a circuit that has resistors connected in parallel. When the script file runs it asks the user first to enter the source voltage and then to enter the resistors' resistance in a vector. The program displays a table with the resistances listed in the first column, the current through the resistor in the second, and the power dissipated in the resistor in the third column. Following the table, the program displays the source current and the total power. Use the script file to solve the following circuit.

11. A truss is a structure made of members jointed at their ends. For the truss shown in the figure, the forces in the nine members are determined by solving the following system of nine equations.

$$\cos(45°)F_1 + F_2 = 0$$

$$F_4 + \cos(48.81°)F_5 - \cos(45°)F_1 = 0$$

$$-\sin(48.81°)F_5 - F_3 - \sin(45°)F_1 = 1000$$

$$\cos(48.81°)F_8 - F_4 = 0,$$

$$-\sin(48.81°)F_8 - F_7 = 500$$

$$F_9 - \cos(48.81°)F_5 - F_6 = 0, \quad F_7 + \sin(48.81°)F_5 = 4000$$

$$\sin(48.81°)F_8 = -1107.14, \quad -\cos(48.81°)F_8 - F_9 = 0$$

Write the equations in a matrix form and use MATLAB to determine the forces in the members. A positive force means tensile force and a negative force means compressive force. Display the results in a table.

12. The graph of the function $f(x) = ax^4 + bx^3 + cx^2 + dx + e$ passes through the points (–3, 6.8), (–1.5, 15.2), (0.5, 14.5), (2, -21.2), and (5, 10). Determine the constants a, b, c, d, and e. (Write a system of five equations with five unknowns and use MATLAB to solve the equations.)

13. During a golf match, a certain number of points are awarded for each eagle and a different number for each birdie. No points are awarded for par, and a certain number of points are deducted for each bogey and a different number deducted for each double bogey (or worse). The newspaper report of an important match neglected to mention what these point values were, but did provide the following table of the results:

Golfer	Eagles	Birdies	Pars	Bogeys	Doubles	Points
Fred	1	5	10	2	0	18
Wilma	2	3	11	1	1	15
Barney	0	3	10	3	2	0
Betty	1	4	10	3	1	12

From the information in the table write four equations in terms of four unknowns. Solve the equations for the unknown points awarded for eagles and birdies and deducted for bogeys and double bogeys.

14. The dissolution of copper sulfide in aqueous nitric acid is described by the following chemical equation:

$$aCuS + bNO_3^- + cH^+ \rightarrow dCu^{2+} + eSO_4^{2-} + fNO + gH_2O$$

where the coefficients a, b, c, d, e, f, and g are the numbers of each molecule participating in the reaction and are unknown. The unknown coefficients are determined by balancing each atom on left and right and then balancing the ionic charge. The resulting equations are:

$$a = d, \quad a = e, \quad b = f, \quad 3b = 4e + f + g, \quad c = 2g, \quad -b + c = 2d - 2e$$

There are seven unknowns and only six equations. A solution can still be obtained, however, by taking advantage of the fact that all the coefficients must be positive integers. Add a 7th equation by guessing $a = 1$ and solve the system of equations. The solution is valid if all the coefficients are positive integers. If this is not the case take $a = 2$ and repeat the solution. Continue the process until all the coefficients in the solution are positive integers.

Chapter 5
Two-Dimensional Plots

Plots are a very useful tool for presenting information. This is true in any field, but especially in science and engineering where MATLAB is mostly used. MATLAB has many commands that can be used for creating different types of plots. These include standard plots with linear axes, plots with logarithmic and semi-logarithmic axes, bar and stairs plots, polar plots, three-dimensional contour surface and mesh plots, and many more. The plots can be formatted to have a desired appearance. The line type (solid, dashed, etc.), color, and thickness can be prescribed, line markers and grid lines can be added, as well as titles and text comments. Several graphs can be plotted in the same plot and several plots can be placed on the same page. When a plot contains several graphs and/or data points, a legend can be added to the plot as well.

This chapter describes how MATLAB can be used to create and format many types of two-dimensional plots. Three-dimensional plots are addressed separately in Chapter 9. An example of a simple two-dimensional plot that was created with MATLAB is shown in Figure 5-1. The figure contains two curves that show the variation of light intensity with distance. One curve is constructed from data points measured in an experiment, and the other curve shows the variation of light as predicted by a theoretical model. The axes in the figure are both linear, and different types of lines (one solid and one dashed) are used for the curves. The theoretical curve is shown with a solid line, while the experimental points are connected with a dashed line. Each data point is marked with a circular marker. The dashed line that connects the experimental points is actually red when the plot is displayed in the Figure Window. As shown, the plot in Figure 5-1 is formatted to have a title, axes' titles, a legend, markers, and a boxed text label.

119

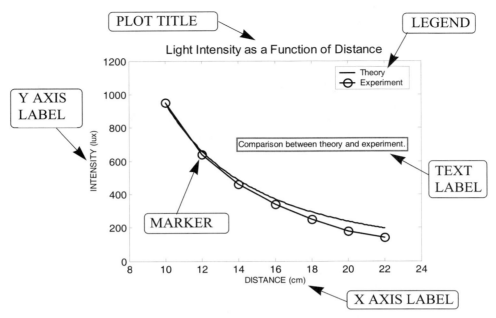

Figure 5-1: Example of a formatted two-dimensional plot.

5.1 THE plot COMMAND

The plot command is used to create two-dimensional plots. The simplest form of the command is:

The arguments x and y are each a vector (one-dimensional array). Both vectors **must** have the same number of elements. When the plot command is executed a figure is created in the Figure Window. If not already open, the Figure Window opens automatically when the command is executed. The figure has a single curve with the x values on the abscissa (horizontal axis), and the y values on the ordinate (vertical axis). The curve is constructed of straight-line segments that connect the points whose coordinates are defined by the elements of the vectors x and y. The vectors, of course, can have any name. The vector that is typed first in the plot command is used for the horizontal axis, and the vector that is typed second is used for the vertical axis.

The figure that is created has axes with linear scale and default range. For example, if a vector *x* has the elements 1, 2, 3, 5, 7, 7.5, 8, 10, and a vector *y* has the elements 2, 6.5, 7, 7, 5.5, 4, 6, 8, a simple plot of *y* versus *x* can be created by typing the following in the Command Window:

```
>> x=[1   2   3   5   7   7.5   8   10];
>> y=[2   6.5   7   7   5.5   4   6   8];
>> plot(x,y)
```

Once the `plot` command is executed, the Figure Window opens and the plot is displayed, as shown in Figure 5-2.

Figure 5-2: The Figure Window with a simple plot.

The plot appears on the screen in blue which is the default line color.

The `plot` command has additional optional arguments that can be used to specify the color and style of the line and the color and type of markers, if any are desired. With these options the command has the form:

Vector Vector   (Optional) Specifiers that   (Optional) Properties with
                  define the type and color   values that can be used to
                  of the line and markers.   specify the line width, and
                                                marker's size and edge, and
                                                fill colors.

Line Specifiers:

Line specifiers are optional and can be used to define the style and color of the line and the type of markers (if markers are desired). The line style specifiers are:

Line Style	Specifier		Line Style	Specifier
solid (default)	-		dotted	:
dashed	--		dash-dot	-.

The line color specifiers are:

Line Color	Specifier
red	r
green	g
blue	b
cyan	c

Line Color	Specifier
magenta	m
yellow	y
black	k
white	w

The marker type specifiers are:

Marker Type	Specifier	Marker Type	Specifier
plus sign	+	square	s
circle	o	diamond	d
asterisk	*	five-pointed star	p
point	.	six-pointed star	h
cross	x	triangle (pointed left)	<
triangle (pointed up)	^	triangle (pointed right)	>
triangle (pointed down)	v		

Notes about using the specifiers:

- The specifiers are typed inside the `plot` command as strings.

- Within the string the specifiers can be typed in any order.

- The specifiers are optional. This means that none, one, two, or all the three can be included in a command.

Some examples:

`plot(x,y)`	A blue solid line connects the points with no markers (default).
`plot(x,y,'r')`	A red solid line connects the points.
`plot(x,y,'--y')`	A yellow dashed line connects the points.
`plot(x,y,'*')`	The points are marked with * (no line between the points).
`plot(x,y,'g:d')`	A green dotted line connects the points that are marked with diamond markers.

`Property Name` and `Property Value`:

Properties are optional and can be used to specify the thickness of the line, the size of the marker, and the colors of the marker's edge line and fill. The Property Name is typed as a string, followed by a comma and a value for the property, all inside the `plot` command.

Four properties and possible values are:

Property Name	Description	Possible Property Values
`LineWidth` (or `linewidth`)	Specifies the width of the line.	A number in units of points (default 0.5).
`MarkerSize` (or `markersize`)	Specifies the size of the marker.	A number in units of points.
`MarkerEdgeColor` (or `markeredgecolor`)	Specifies the color of the marker, or the color of the edge line for filled markers.	Color specifiers from the table above, typed as a string.
`MarkerFaceColor` (or `markerfacecolor`)	Specifies the color of the filling for filled markers.	Color specifiers from the table above, typed as a string.

For example, the command:

```
plot(x,y,'-mo','LineWidth',2,'markersize',12,
     'MarkerEdgeColor','g','markerfacecolor','y')
```

creates a plot that connects the points with a magenta solid line and circles as markers at the points. The line width is two points and the size of the circle markers is 12 points. The markers have a green edge line and yellow filling.

A note about line specifiers and properties:

The three line specifiers, which are the style and color of the line, and the type of the marker can also be assigned with a `PropertyName` argument followed by a `PropertyValue` argument. The Property Names for the line specifiers are:

Specifier	Property Name	Possible Property Values
Line Style	`linestyle` (or `LineStyle`)	Line style specifier from the table above, typed as a string.
Line Color	`color` (or `Color`)	Color specifiers from the table above, typed as a string.
Marker	`marker` (or `Marker`)	Marker specifier from the table above, typed as a string.

As with any command, the `plot` command can be typed in the Command Window, or it can be included in a script file. It also can be used in a function file (explained in Chapter 6). It should also be remembered that before the `plot` command can be executed the vectors x and y must have assigned elements. This can be done, as was explained in Chapter 2, by entering values directly, by using commands, or as the result of mathematical operations. The next two subsections show examples of creating simple plots.

5.1.1 Plot of Given Data

In this case given data is first used to create vectors that are then used in the `plot` command. For example, the following table contains sales data of a company from 1988 to 1994.

YEAR	1988	1989	1990	1991	1992	1993	1994
SALES (millions)	8	12	20	22	18	24	27

To plot this data, the list of years is assigned to one vector (named `yr`), and the corresponding sale data is assigned to a second vector (named `sle`). The Command Window where the vectors are created and the `plot` command is used is shown below:

```
>> yr=[1988:1:1994];
>> sle=[8  12  20  22  18  24  27];
>> plot(yr,sle,'--r*','linewidth',2,'markersize',12)
>>
```

Line Specifiers: dashed red line and asterisk marker.

Property Name and Property Value: the line width is 2 points and the markers size is 12 point.

Once the `plot` command is executed the Figure Window with the plot, as shown in Figure 5-3, opens. The plot appears on the screen in red.

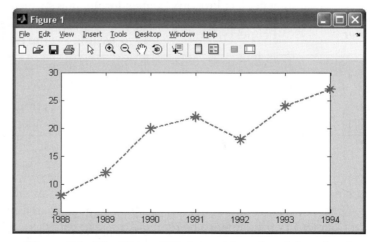

Figure 5-3: The Figure Window with a plot of the sales data.

5.1.2 *Plot of a Function*

In many situations there is a need to plot a given function. This can be done in MATLAB by using the `plot` or the `fplot` commands. The use of the `plot` command is explained below. The `fplot` command is explained in detail in the next section.

In order to plot a function $y = f(x)$ with the `plot` command, the user needs to first create a vector of values of x for the domain that the function will be plotted. Then, a vector y is created with the corresponding values of $f(x)$ by using element-by-element calculations (see Chapter 3). Once the two vectors exist, they can be used in the `plot` command.

As an example, the `plot` command is used to plot the function $y = 3.5^{-0.5x} \cos(6x)$ for $-2 \le x \le 4$. A program that plots this function is shown in the following script file.

```
%  A script file that creates a plot of
%  the function: 3.5.^(-0.5*x).*cos(6x)
x=[-2:0.01:4];          Create vector x with the domain of the function.

y=3.5.^(-0.5*x).*cos(6*x);
                         Create vector y with the function
                         value at each x.

plot(x,y)                Plot y as a function of x.
```

Once the script file is executed, the plot is created in the Figure Window, as shown in Figure 5-4. Since the plot is made up of segments of straight lines that connect the points, to obtain an accurate plot of a function, the spacing between the elements of the vector x must be appropriate. Smaller spacing is needed for a func-

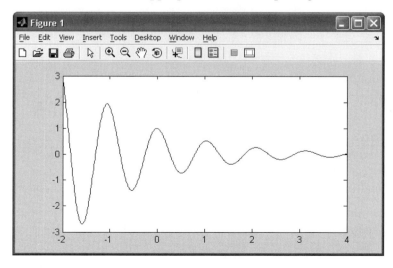

Figure 5-4:   The Figure Window with a plot of the function: $y = 3.5^{-0.5x} \cos(6x)$.

tion that changes rapidly. In the last example a small spacing of 0.01 produced the plot that is shown in Figure 5-4. However, if the same function in the same domain is plotted with much larger spacing, for example 0.3, the plot that is obtained, shown in Figure 5-5, gives a distorted picture of the function. Note also

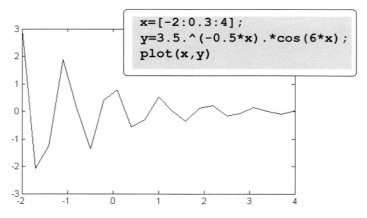

```
x=[-2:0.3:4];
y=3.5.^(-0.5*x).*cos(6*x);
plot(x,y)
```

Figure 5-5: A plot of the function $y = 3.5^{-0.5x}\cos(6x)$ **with large spacing.**

that in Figure 5-4 the plot is shown with the Figure Window, while in Figure 5-5, only the plot is shown. The plot can be copied from the Figure Window (in the **Edit** menu select **Copy Figure**) and then pasted into other applications.

5.2 THE fplot COMMAND

The fplot command plots a function with the form $y = f(x)$ between specified limits. The command has the form:

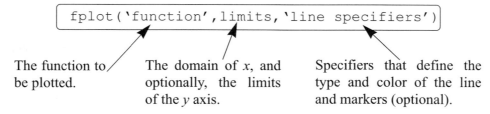

$$\boxed{\texttt{fplot('function',limits,'line specifiers')}}$$

The function to be plotted.   The domain of x, and optionally, the limits of the y axis.   Specifiers that define the type and color of the line and markers (optional).

'function': The function can be typed directly as a string inside the command. For example, if the function that is being plotted is $f(x) = 8x^2 + 5\cos(x)$, it is typed as: `'8*x^2+5*cos(x)'`. The function can include MATLAB built-in functions and functions that are created by the user (covered in Chapter 6).

- The function to be plotted can be typed as a function of any letter. For example, the function in the previous paragraph can be typed as: `'8*z^2+5*cos(z)'`, or `'8*t^2+5*cos(t)'`.

- The function can not include previously defined variables. For example, in the function above it is not possible to assign 8 to a variable, and then use the variable when the function is typed in the `fplot` command.

limits: The limits is a vector with two elements that specify the domain of x [xmin, xmax], or a vector with four elements that specifies the domain of x and the limits of the y-axis [xmin, xmax, ymin, ymax].

Line specifiers: The line specifiers are the same as in the plot command. For example, a plot of the function $y = x^2 + 4\sin(2x) - 1$ for $-3 \le x \le 3$ can be created with the `fplot` command by typing:

```
>> fplot('x^2+4*sin(2*x)-1',[-3 3])
```

in the Command Window. The figure that is obtained in the Figure Window is shown in Figure 5-6.

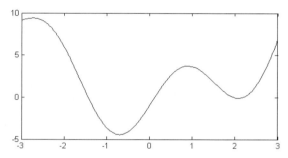

Figure 5-6: A plot of the function $y = x^2 + 4\sin(2x) - 1$.

5.3 PLOTTING MULTIPLE GRAPHS IN THE SAME PLOT

In many situations there is a need to make several graphs in the same plot. This is shown, for example, in Figure 5-1 where two graphs are plotted in the same figure. There are three methods to plot multiple graphs in one figure. One is by using the `plot` command, the other is by using the `hold on`, `hold off` commands, and the third is by using the `line` command.

5.3.1 Using the `plot` Command

Two or more graphs can be created in the same plot by typing pairs of vectors inside the `plot` command. The command:

$$plot(x, y, u, v, t, h)$$

creates three graphs: y vs. x, v vs. u, and h vs. t, all in the same plot. The vectors of each pair must be of the same length. MATLAB automatically plots the graphs in different colors so that they can be identified. It is also possible to add line specifiers following each pair. For example the command:

$$plot(x, y, `-b', u, v, `--r', t, h, `g:')$$

plots y vs. x with a solid blue line, v vs. u with a dashed red line, and h vs. t with a dotted green line.

Sample Problem 5-1: Plotting a function and its derivatives

Plot the function $y = 3x^3 - 26x + 10$, and its first and second derivatives, for $-2 \leq x \leq 4$, all in the same plot.

Solution

The first derivative of the function is: $y' = 9x^2 - 26$.

The second derivative of the function is: $y'' = 18x$.

A script file that creates a vector x, and calculates the values of y, y', and y'' is:

```
x=[-2:0.01:4];        Create vector x with the domain of the function.
y=3*x.^3-26*x+6;      Create vector y with the function value at each x.
yd=9*x.^2-26;         Create vector yd with values of the first derivative.
ydd=18*x;             Create vector ydd with values of the second derivative.
plot(x,y,'-b',x,yd,'--r',x,ydd,':k')

          Create three graphs, y vs. x, yd vs. x, and ydd vs. x in the same figure.
```

The plot that is created is shown in Figure 5-7.

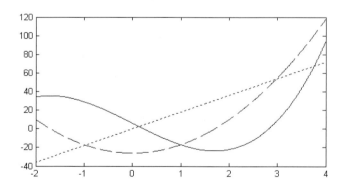

Figure 5-7: A plot of the function $y = 3x^3 - 26x + 10$ and its first and second derivatives.

5.3.2 Using the hold on, hold off Commands

To plot several graphs using the hold on, hold off commands, one graph is plotted first with the plot command. Then the hold on command is typed. This keeps the Figure Window with the first plot open, including the axis properties and formatting (see Section 5.4) if any was done. Additional graphs can be added with `plot` commands that are typed next. Each `plot` command creates a graph that is added to that figure. The `hold off` command stops this process. It returns MATLAB to the default mode in which the `plot` command erases the previous plot and resets the axis properties.

As an example, a solution of Sample Problem 5-1 using the `hold on`, `hold off` commands, is shown in the following script file:

```
x=[-2:0.01:4];
y=3*x.^3-26*x+6;
yd=9*x.^2-26;
ydd=18*x;
plot(x,y,'-b')          The first graph is created.
hold on
plot(x,yd,'--r')        Two more graphs are added to the figure.
plot(x,ydd,':k')
hold off
```

5.3.3 Using the `line` Command

With the `line` command additional graphs (lines) can be added to a plot that already exists. The form of the line command is:

line(x,y,'PropertyName',PropertyValue)

(Optional) Properties with values that can be used to specify the line style, color, and width, marker type, size, and edge and fill colors.

The format of the `line` command is almost the same as the `plot` command (see Section 5.1). The `line` command does not have the line specifiers, but the line style, color, and marker can be specified with the Property Name and property value features. The properties are optional and if none are entered MATLAB uses default properties and values. For example, the command:

line(x,y,'linestyle','--','color','r','marker','o')

will add a dashed red line with circular markers to a plot that already exists.

The major difference between the `plot` and `line` commands is that the `plot` command starts a new plot every time it is executed, while the `line` command adds lines to a plot that already exists. To make a plot that has several graphs, a plot command is typed first and then line commands are typed for additional graphs. (If a line command is entered before a plot command an error message is displayed.)

The solution to Sample Problem 5-1, which is the plot in Figure 5-7, can be obtained by using the `plot` and `line` commands as shown in the following script file:

```
x=[-2:0.01:4];
y=3*x.^3-26*x+6;
yd=9*x.^2-26;
ydd=18*x;
plot(x,y,'LineStyle','-','color','b')
line(x,yd,'LineStyle','--','color','r')
line(x,ydd,'linestyle',':','color','k')
```

5.4 FORMATTING A PLOT

The `plot` and `fplot` commands create bare plots. Usually, however, a figure that contains a plot needs to be formatted to have a specific look and to display information in addition to the graph itself. It can include specifying axis labels, plot title, legend, grid, range of custom axis, and text labels.

Plots can be formatted by using MATLAB commands that follow the `plot` or `fplot` commands, or interactively by using the plot editor in the Figure Window. The first method is useful when a `plot` command is a part of a computer program (script file). When the formatting commands are included in the program, a formatted plot is created every time the program is executed. On the other hand, formatting that is done in the Figure Window with the plot editor after a plot has been created holds only for that specific plot, and will have to be repeated the next time the plot is created.

5.4.1 Formatting a Plot Using Commands

The formatting commands are entered after the `plot` or the `fplot` commands. The various formatting commands are:

The `xlabel` **and** `ylabel` **commands:**

Labels can be placed next to the axes with the `xlabel` and `ylabel` commands which have the form:

```
xlabel('text as string')
ylabel('text as string')
```

The `title` **command:**

A title can be added to the plot with the command:

```
title('text as string')
```

The text is placed at the top of the figure as a title.

The `text` command:

A text label can be placed in the plot with the `text` or `gtext` commands:

```
text(x,y,'text as string')
gtext('text as string')
```

The `text` command places the text in the figure such that the first character is positioned at the point with the coordinates `x`, `y` (according to the axes of the figure). The `gtext` command places the text at a position specified by the user. When the command is executed, the Figure Window opens and the user specifies the position with the mouse.

The `legend` command:

The `legend` command places a legend on the plot. The legend shows a sample of the line type of each graph that is plotted, and places a label, specified by the user, beside the line sample. The form of the command is:

```
legend('string1','string2', ..... ,pos)
```

The strings are the labels that are placed next to the line sample. Their order corresponds to the order that the graphs were created. The `pos` is an optional number that specifies where in the figure the legend is placed. The options are:

pos = -1   Places the legend outside the axes boundaries on the right side.
pos = 0   Places the legend inside the axes boundaries in a location that interferes the least with the graphs.
pos = 1   Places the legend at the upper-right corner of the plot (default).
pos = 2   Places the legend at the upper-left corner of the plot.
pos = 3   Places the legend at the lower-left corner of the plot.
pos = 4   Places the legend at the lower-right corner of the plot.

Formatting the text in the `xlabel`, `ylabel`, `title`, `text`

and `legend` **commands:**

The text in the string that is included in the commands and is displayed when the commands are executed can be formatted. The formatting can be used to define the font, size, position (superscript, subscript), style (italic, bold, etc.), and color of the characters, color of the background, and to define many other details of the display. Some of the more common formatting possibilities are described below. A complete explanation of all the formatting features can be found in the Help Window under Text and Text Properties. The formatting can be done either by adding modifiers inside the string, or by adding to the command optional `PropertyName` and `PropertyValue` arguments following the string.

The modifiers are characters that are inserted within the string. Some of the modifiers that can be added are:

Modifier	Effect
`\bf`	bold font.
`\it`	italic style.
`\rm`	normal font.

Modifier	Effect
`\fontname{fontname}`	specified font is used.
`\fontsize{fontsize}`	specified font size is used.

These modifiers affect the text from the point that they are inserted until the end of the string. It is also possible to have the modifiers applied to only a section of the string by typing the modifier and the text to be affected inside braces { }.

Subscript and superscript:

A single character can be displayed as a subscript or a superscript by typing _ (the underscore character) or ^ in front of the character, respectively. Several consecutive characters can be displayed as subscript or a superscript by typing the characters inside braces { } following the _ or the ^.

Greek characters:

Greek characters can be included in the text by typing `\name of the letter` within the string. To display a lowercase Greek letter the name of the letter should be typed in all lowercase English characters, To display a capital Greek letter the name of the letter should start with a capital letter. Some examples are:

Characters in the string	Greek Letter
`\alpha`	α
`\beta`	β
`\gamma`	γ
`\theta`	θ
`\pi`	π
`\sigma`	σ

Characters in the string	Greek Letter
`\Phi`	Φ
`\Delta`	Δ
`\Gamma`	Γ
`\Lambda`	Λ
`\Omega`	Ω
`\Sigma`	Σ

Formatting of the text that is displayed by the `xlabel`, `ylabel`, `title`, and `text` commands can also be done by adding optional `PropertyName` and `PropertyValue` arguments following the string inside the command. With this option the `text` command, for example, has the form:

```
text(x,y,'text as string',PropertyName,PropertyValue)
```

In the other three commands the `PropertyName` and `PropertyValue` arguments are added in the same way. The `PropertyName` is typed as a string, and the `PropertyValue` is typed as a number if the property value is a number and as a string if the property value is a word or a letter character. Some of the Property Names and corresponding possible Property Values are:

Property Name	Description	Possible Property Values
Rotation	Specifies the orientation of the text.	Scalar (degrees) Default: 0
FontAngle	Specifies italic or normal style characters.	normal, italic Default: normal
FontName	Specifies the font for the text.	Font name that is available in the system.
FontSize	Specifies the size of the font.	Scalar (points) Default: 10
FontWeight	Specifies the weight of the characters.	light, normal, bold Default: normal
Color	Specifies the color of the text.	Color specifiers (See Section 5.1).
Background-Color	Specifies the background color (rectangular area).	Color specifiers (See Section 5.1).
EdgeColor	Specifies the color of the edge of a rectangular box around the text.	Color specifiers (See Section 5.1). Default: none.
LineWidth	Specifies the width of the edge of a rectangular box around the text.	Scalar (points) Default: 0.5

The `axis` command:

When the `plot(x,y)` command is executed, MATLAB creates axes with limits that are based on the minimum and maximum values of the elements of `x` and `y`. The `axis` command can be used to change the range and the appearance of the axes. In many situations a graph looks better if the range of the axes extend beyond the range of the data. The following are some of the possible forms of the axis command:

axis([xmin,xmax,ymin,ymax])   Sets the limits of both the x and y axes (xmin, xmax, ymin, and ymax are numbers).

axis equal   Sets the same scale for both axes.

axis square   Sets the axes region to be square.

axis tight   Sets the axis limits to the range of the data.

The grid command:

grid on   Adds grid lines to the plot.

grid off   Removes grid lines from the plot.

An example of formatting a plot by using commands is given in the following script file that was used to generate the formatted plot in Figure 5-1.

```
x=[10:0.1:22];
y=95000./x.^2;
xd=[10:2:22];
yd=[950  640  460  340  250  180  140];
plot(x,y,'-','LineWidth',1.0)
xlabel('DISTANCE (cm)')
ylabel('INTENSITY (lux)')
title('\fontname{Arial}Light Intensity as a Function of Distance','FontSize',14)
axis([8 24 0 1200])
text(14,700,'Comparison between theory and experiment.','EdgeColor','r','LineWidth',2)
hold on
plot(xd,yd,'ro--','linewidth',1.0,'markersize',10)
legend('Theory','Experiment',0)
hold off
```

Formatting text inside the title command.

Formatting text inside the text command.

5.4.2 Formatting a Plot Using the Plot Editor

A plot can be formatted interactively in the Figure Window by clicking on the plot and/or using the menus. Figure 5-8 shows the Figure Window with the plot of Figure 5-1. The Plot Editor can be used to introduce new formatting items, or to modify formatting that was initially introduced with the formatting commands.

Click the arrow button to start the plot edit mode. Then click on an item. A window with formatting tool for the item opens.

Use the **Edit** and **Insert** menus to add formatting objects, or to edit existing objects.

Change position of labels, legends and other objects by clicking on the object and dragging.

Figure 5-8: Formatting a plot using the plot editor.

5.5 *PLOTS WITH LOGARITHMIC AXES*

Many science and engineering applications require plots in which one or both axes have a logarithmic (log) scale. Log scales provide means for presenting data over a wide range of values. It also provides a tool for identifying characteristics of data and possible forms of mathematical relationships that can be appropriate for modeling the data (see Section 8.2.2).

MATLAB commands for making plots with log axes are:

semilogy(x,y)   Plots y versus x with a log (base 10) scale for the y axis and linear scale for the x axis.

semilogx(x,y)   Plots y versus x with a log (base 10) scale for the x axis and linear scale for the y axis.

loglog(x,y)   Plots y versus x with a log (base 10) scale for both axes.

Line specifiers and Property Name and property value can be added to the commands (optional) just as in the plot command. As an example, Figure 5-9 shows a plot of the function $y = 2^{(-0.2x + 10)}$ for $0.1 \le x \le 60$. The figure shows four plots of the same function: one with linear axes, one with log scale for the y-axis, one with log scale for the x-axis, and one with log scale on both axes.

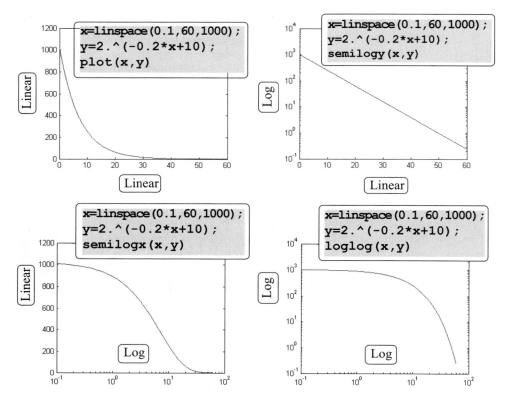

Figure 5-9: Plots of $y = 2^{(-0.2x + 10)}$ with linear, semilog, and loglog scales.

Notes for plots with logarithmic axes:

- The number zero cannot be plotted on a log scale (since a log of zero is not defined).

- Negative numbers cannot be plotted on log scales (since a log of a negative number is not defined).

5.6 PLOTS WITH ERROR BARS

Experimental data that is measured and then displayed in plots frequently contains error and scatter. Even data that is generated by computational models includes error or uncertainty that depends on the accuracy of the input parameters and the assumptions in the mathematical models that are used. One method of plotting data that displays the error, or uncertainty, is by using error bars. An error bar is typically a short vertical line that is attached to a data point in a plot. It shows the magnitude of the error that is associated to the value that is displayed by the data point. For example, Figure 5-10 shows a plot with error bars of the experimental data from Figure 5-1.

Figure 5-10: A plot with error bars.

Plots with error bars can be done in MATLAB with the `errorbar` command. Two forms of the command, one for making plots with symmetric error bars (with respect to the value of the data point), and the other for nonsymmetric error bars, at each point are presented. When the error is symmetric, the error bar extend the same length above and below the data point and the command has the form:

Vectors with horizontal and vertical coordinates of each point.   Vector with the values of the error at each point.

- The length of the three vectors x, y, and e must be the same.

- The length of the error bars is twice the value of e. At each point the error bar extends from `y(i)-e(i)` to `y(i)+e(i)`.

The plot in Figure 5-10, which has symmetric error bars, was done by executing the following code:

```
xd=[10:2:22];
yd=[950 640 460 340 250 180 140];
ydErr=[30 20 18 35 20 30 10]
errorbar(xd,yd,ydErr)
xlabel('DISTANCE (cm)')
ylabel('INTENSITY (lux)')
```

The command for making a plot with error bars that are not symmetric is:

Vectors with horizontal and vertical coordinates of each point.   Vector with the upper-bound values of the error at each point.   Vector with the lower-bound values of the error at each point.

- The length of the three vectors x, y, d, and u must be the same.

- At each point the error bar extend from y(i)-d(i) to y(i)+u(i).

5.7 PLOTS WITH SPECIAL GRAPHICS

All the plots that have been presented so far in this chapter are line plots in which the data points are connected by lines. In many situations plots with different graphics or geometry can present data more effectively. MATLAB has many options for creating a wide variety of plots. These include bar, stairs, stem, pie plots and many more. The following shows some of the special graphics plots that can be created with MATLAB. A complete list of the plotting functions that MATLAB has and information on how to use them can be found in the Help Window. In this window first choose "Functions by Category,", then select "Graphics" and then select "Basic Plots and Graphs" or "Specialized Plotting."

Bar (vertical and horizontal), stairs, and stem plots are presented in the following table using the sales data from Section 5.1.1.

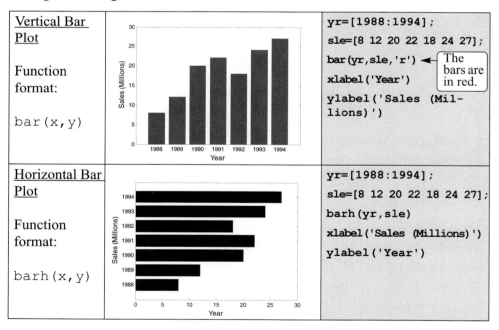

Vertical Bar Plot
 	
Function format:
 	
bar(x,y)
		`yr=[1988:1994];`
		
		`sle=[8 12 20 22 18 24 27];`
		
		`bar(yr,sle,'r')` ← The bars are in red.
		
		`xlabel('Year')`
		
		`ylabel('Sales (Millions)')`

Horizontal Bar Plot
 	
Function format:
 	
barh(x,y)
		`yr=[1988:1994];`
		
		`sle=[8 12 20 22 18 24 27];`
		
		`barh(yr,sle)`
		
		`xlabel('Sales (Millions)')`
		
		`ylabel('Year')`

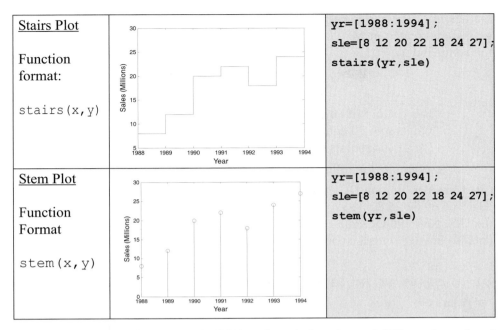

| Stairs Plot

Function format:

`stairs(x,y)` | | `yr=[1988:1994];`
`sle=[8 12 20 22 18 24 27];`
`stairs(yr,sle)` |
| Stem Plot

Function Format

`stem(x,y)` | | `yr=[1988:1994];`
`sle=[8 12 20 22 18 24 27];`
`stem(yr,sle)` |

Pie charts are useful for visualizing the relative sizes of different but related quantities. For example, the table below shows the grades that were assigned to a class. The data below is used to create the pie chart that follows.

Grade	A	B	C	D	E
Number of Students	11	18	26	9	5

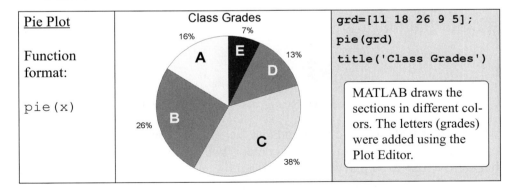

| Pie Plot

Function format:

`pie(x)` | | `grd=[11 18 26 9 5];`
`pie(grd)`
`title('Class Grades')`

MATLAB draws the sections in different colors. The letters (grades) were added using the Plot Editor. |

5.8 HISTOGRAMS

Histograms are plots that show the distribution of data. The overall range of a given set of data points is divided to smaller subranges (bins), and the histogram shows how many data points are in each bin. The histogram is a vertical bar plot in which the width of each bar is equal to the range of the corresponding bin, and the height of the bar corresponds to the number of data points in the bin. Histograms are created in MATLAB with the `hist` command. The simplest form of the command is:

$$\boxed{\texttt{hist(y)}}$$

y   is a vector with the data points. MATLAB divides the range of the data points into 10 equally spaced subranges (bins), and then plots the number of data points in each bin.

For example, the following data points are the daily maximum temperature (in °F) in Washington DC during the month of April, 2002: 58 73 73 53 50 48 56 73 73 66 69 63 74 82 84 91 93 89 91 80 59 69 56 64 63 66 64 74 63 69, (data from the U.S. National Oceanic and Atmospheric Administration). A histogram of this data is obtained with the commands:

```
>> y=[58 73 73 53 50 48 56 73 73 66 69 63 74 82 84 91 93 89
91 80 59 69 56 64 63 66 64 74 63 69];
>> hist(y)
```

The plot that is generated is shown in Figure 5-10 (the axes titles were added using the Plot Editor). The smallest value in the data set is 48 and the largest is 93,

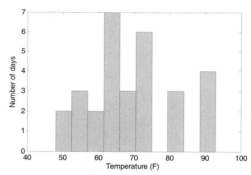

Figure 5-11: Histogram of temperature data.

which means that the range is 45 and that the width of each bin is 4.5. The range of the first bin is from 48 to 52.5 and contains two points. The range of the second bin is from 52.5 to 57 and contains three points, and so on. Two of the bins (75 to 79.5 and 84 to 88.5) do not contain any points.

Since the division of the data range into 10 equally spaced bins might not be the division that is preferred by the user, the number of bins can be defined by the user to be different than 10. This can be done either by specifying the number of bins, or by specifying the center point of each bin as shown in the following two forms of the `hist` command:

> hist(y,nbins)   or   hist(y,x)

nbins   is a scalar that defines the number of bins. MATLAB divides the range to equally spaced sub-ranges.

x   is a vector that specifies the location of the center of each bin (the distance between the centers does not have to be the same for all the bins). The edges of the bins are at the middle point between the centers.

In the example above the user might prefer to divide the temperature range into 3 bins. This can be done with the command:

```
>> hist(y,3)
```

As shown on the right, the histogram that is generated has three equally spaced bins.

The number and width of the bins can also be specified by a vector x whose elements define the centers of the bins. For example, shown on the right is a histogram that displays the temperature data from above in 6 bins with an equal width of 10 degrees. The elements of the vector x for this plot are 45, 55, 65, 75, 85, and 95. The plot was obtained with the following commands:

```
>> x=[45:10:95]
x =
    45    55    65    75    85    95
>> hist(y,x)
```

The `hist` command can be used with options that provide numerical output in addition to plotting a histogram. An output of the number of data points in each bin can be obtained with one of the following commands:

> n=hist(y)   n=hist(y,nbins)   n=hist(y,x)

The output n is a vector. The number of elements in n is equal to the number of bins and the value of each element of n is the number of data points (frequency count) in the corresponding bin. For example, the histogram in Figure 5-10 can also be created with the following command:

```
>> n = hist(y)
n =
    2   3   2   7   3   6   0   3   0   4
```

The vector n shows how may elements are in each bin.

The vector n shows that the first bin has 2 data points, the second bin has 3 data points, and so on.

An additional optional numerical output is the location of the bins. This output can be obtained with one of the following commands:

[n xout]=hist(y)   [n xout]=hist(y,nbins)

xout is a vector in which the value of each element is the location of the center of the corresponding bin. For example, for the histogram in Figure 5-10:

```
>> [n xout]=hist(y)
n =
    2    3    2    7    3    6    0    3    0    4
xout =
   50.2500   54.7500   59.2500   63.7500   68.2500   72.7500
   77.2500   81.7500   86.2500   90.7500
```

The vector xout shows that the center of the first bin is at 50.25, the center of the second bin is at 54.75 and so on.

5.9 POLAR PLOTS

Polar coordinates, in which the position of a point in a plane is defined by the angle θ and the radius (distance) to the point, are frequently used in the solution of science and engineering problems. The polar command is used to plot functions in polar coordinates. The command has the form:

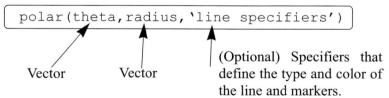

polar(theta,radius,'line specifiers')

Vector   Vector   (Optional) Specifiers that define the type and color of the line and markers.

where theta and radius are vectors whose elements define the coordinates of the points to be plotted. The polar command plots the points and draws the polar grid. The line specifiers are the same as in the plot command. To plot a function $r = f(\theta)$ in a certain domain, a vector for values of θ is created first, and then a vector r with the corresponding values of $f(\theta)$ is created using element-by-element calculations. The two vectors are then used in the polar command.

For example, a plot of the function $r = 3\cos^2(0.5\theta) + \theta$ for $0 \le \theta \le 2\pi$ is shown below.

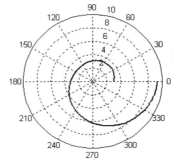

```
t=linspace(0,2*pi,200);
r=3*cos(0.5*t).^2+t;
polar(t,r)
```

5.10 PLOTTING MULTIPLE PLOTS ON THE SAME PAGE

Multiple plots on the same page can be created with the subplot command, which has the form:

$$\boxed{\texttt{subplot(m,n,p)}}$$

The command divides the Figure Window (page when printed) into $m \times n$ rectangular subplots where plots will be created. The subplots are arranged like elements in a $m \times n$ matrix where each element is a subplot. The subplots are numbered from 1 through $m \cdot n$. The upper left is 1 and the lower right is the number $m \cdot n$. The numbers increase from left to right within a row, from the first row to the last. The command subplot(m,n,p) makes the subplot p current. This means that the next plot command (and any formatting commands) will create a plot (with the corresponding format) in this subplot. For example, the command subplot(3,2,1) creates 6 areas arranged in 3 rows and 2 columns as shown, and makes the upper left subplot current. An example of using the subplot command is shown in the solution of Sample Problem 5-2.

(3,2,1)	(3,2,2)
(3,2,3)	(3,2,4)
(3,2,5)	(3,2,6)

5.11 MULTIPLE FIGURE WINDOWS

When the plot or any other command that generates a plot is executed, the Figure Window opens (if not already open) and displays the plot. MATLAB labels the Figure Window as Figure 1 (see the top left corner of the Figure Window that is displayed in Figure 5-4). If the Figure Window is already open when the plot or any other command that makes a plot is executed, a new plot is displayed in the

Figure Window that is already open (replacing the existing plot). Commands that format plots are applied to the plot in the Figure Window that is open.

It is possible, however, to open additional Figure Windows and have several of them open (with plots) at the same time. This is done by typing the command `figure`. Every time the command `figure` is entered, MATLAB opens a new Figure Window. If a command that creates a plot is entered after a `figure` command, MATLAB generates and displays the new plot in the last Figure Window that was opened, which is called the active or current window. MATLAB labels the new Figure Windows successively; i.e., Figure 2, Figure 3, and so on. For example, after the following three commands are entered, two Figure Windows that are shown in Figure 5-11 are displayed.

```
>> fplot('x*cos(x)',[0,10])        Plot displayed in Figure 1 Window.
>> figure                           Figure 2 Window opens.
>> fplot('exp(-0.2*x)*cos(x)',[0,10])   Plot displayed in Figure 2 Window.
```

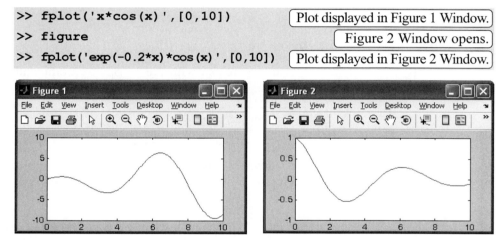

Figure 5-12: Histogram of temperature data.

The `figure` command can also have an input argument that is a number (integer) `figure(n)`. The number corresponds to the number of a corresponding Figure Window. When the command is executed, Figure Window number n becomes the active Figure Window (if a Figure Window with this number does not exist, a new window with this number opens). When commands that create new plots are executed, the plots that they generate are displayed in the active Figure Window. In the same way, commands that format plots are applied to the plot in the active window. The `figure(n)` command provides means for having a program in a script file that open and make plots in a few defined Figure Windows. (If several `figure` commands are used in a program instead, new Figure Windows will open every time the script file is executed.)

Figure Windows can be closed with the `close` command. Several forms of the command are:

`close` closes the active Figure Window.

`close(n)` closes the nth Figure Window.

`close all` closes all Figure Windows that are open.

5.12 *EXAMPLES OF MATLAB APPLICATIONS*

Sample Problem 5-2:   Piston-crank mechanism

The piston-connecting rod-crank mechanism is used in many engineering applications. In the mechanism shown in the following figure, the crank is rotating at a constant speed of 500 rpm.

Calculate and plot the position, velocity, and acceleration of the piston for one revolution of the crank. Make the three plots on the same page. Set $\theta = 0°$ when $t = 0$.

Solution

The crank is rotating with a constant angular velocity $\dot{\theta}$. This means that if we set $\theta = 0°$ when $t = 0$, then at time t the angle θ is given by $\theta = \dot{\theta}t$, and that $\ddot{\theta} = 0$ at all times.

The distances d_1 and h are given by:

$$d_1 = r\cos\theta \quad \text{and} \quad h = r\sin\theta$$

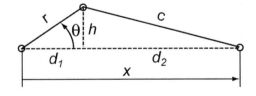

Knowing h, the distance d_2 can be calculated using the Pythagorean theorem:

$$d_2 = (c^2 - h^2)^{1/2} = (c^2 - r^2\sin^2\theta)^{1/2}$$

The position x of the piston is then given by:

$$x = d_1 + d_2 = r\cos\theta + (c^2 - r^2\sin^2\theta)^{1/2}$$

The derivative of x with respect to time gives the velocity of the piston:

$$\dot{x} = -r\dot{\theta}\sin\theta - \frac{r^2\dot{\theta}\sin 2\theta}{2(c^2 - r^2\sin^2\theta)^{1/2}}$$

The second derivative of x with respect to time gives the acceleration of the piston:

$$\ddot{x} = -r\dot{\theta}^2\cos\theta - \frac{4r^2\dot{\theta}^2\cos 2\theta(c^2 - r^2\sin^2\theta) + (r^2\dot{\theta}\sin 2\theta)^2}{4(c^2 - r^2\sin^2\theta)^{3/2}}$$

In the equation above $\dot{\theta}$ was taken to be zero.

A MATLAB program (script file) that calculates and plots the position, velocity, and acceleration of the piston for one revolution of the crank is shown below:

```
THDrpm=500; r=0.12; c=0.25;              Define θ, r and c.
THD=THDrpm*2*pi/60;              Change the units of θ̇ from rpm to rad/s.
tf=2*pi/THD;              Calculate the time for one revolution of the crank.
t=linspace(0,tf,200);              Create a vector for the time with 200 elements.
TH=THD*t;              Calculate θ for each t.
d2s=c^2-r^2*sin(TH).^2;              Calculate d₂ squared for each θ.
x=r*cos(TH)+sqrt(d2s);              Calculate x for each θ.
xd=-r*THD*sin(TH)-(r^2*THD*sin(2*TH))./(2*sqrt(d2s));
xdd=-r*THD^2*cos(TH)-(4*r^2*THD^2*cos(2*TH).*d2s+
(r^2*sin(2*TH)*THD).^2)./(4*d2s.^(3/2));
subplot(3,1,1)              Calculate ẋ and ẍ for each θ.
plot(t,x)              Plot x vs. t.
grid              Format the first plot.
xlabel('Time (s)')
ylabel('Position (m)')
subplot(3,1,2)
plot(t,xd)              Plot ẋ vs. t.
grid              Format the second plot.
xlabel('Time (s)')
ylabel('Velocity (m/s)')
subplot(3,1,3)
plot(t,xdd)              Plot ẍ vs. t.
grid              Format the third plot.
xlabel('Time (s)')
ylabel('Acceleration (m/s^2)')
```

When the script file runs it generates the three plots on the same page as shown in Figure 5-11. The figure nicely shows that the velocity of the piston is zero at the end points of the travel range where the piston changes the direction of the motion. The acceleration is maximum (directed to the left) when the piston is at the right end.

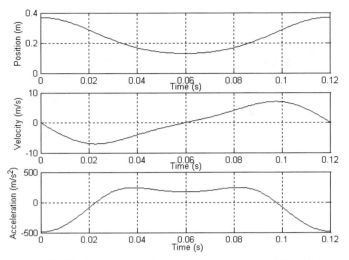

Figure 5-13: Position, velocity, and acceleration of the piston vs. time.

Sample Problem 5-3: Electric Dipole

The electric field at a point due to a charge is a vector **E** with magnitude E given by Coulomb's law:

$$E = \frac{1}{4\pi\varepsilon_0}\frac{q}{r^2}$$

where $\varepsilon_0 = 8.8541878 \times 10^{-12}\frac{C^2}{N \cdot m^2}$ is the permittivity constant, q is the magnitude of the charge, and r is the distance between the charge and the point. The direction of **E** is along the line that connects the charge with the point. **E** points outward from q if q is positive, and toward q if q is negative. An electric dipole is created when a positive charge and a negative charge of equal magnitude are placed some distance apart. The electric field, **E**, at any point is obtained by superposition of the electric field of each charge.

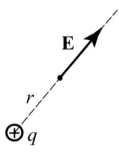

An electric dipole with $q = 12 \times 10^{-9}C$ is created, as shown in the figure. Determine and plot the magnitude of the electric field along the x-axis from $x = -5cm$ to $x = 5cm$.

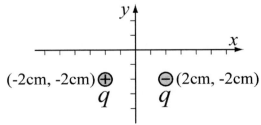

Solution

The electric field **E** at any point $(x, 0)$ along the x-axis is obtained by adding the electric field vectors due to each of the charges.

$$\mathbf{E} = \mathbf{E}_- + \mathbf{E}_+$$

The magnitude of the electric field is the length of the vector **E**.

     The problem is solved by following these steps:

Step 1:   Create a vector x for points along the x-axis.

Step 2:   Calculate the distance (and distance^2) from each charge to the points on the x-axis.

$$r_{minus} = \sqrt{(0.02-x)^2 + 0.02^2} \qquad r_{plus} = \sqrt{(x+0.02x)^2 + 0.02^2}$$

Step 3:   Write unit vectors in the direction from each charge to the points on the x-axis.

$$\mathbf{E}_{minusUV} = \frac{1}{r_{minus}}((0.02-x)\mathbf{i} - 0.02\mathbf{j})$$

$$\mathbf{E}_{plusUV} = \frac{1}{r_{plus}}((x+0.02)\mathbf{i} + 0.02\mathbf{j})$$

Step 4:   Calculate the magnitude of the vector $\mathbf{E}_-$ and $\mathbf{E}_+$ at each point by using Coulomb's law.

$$E_{minusMAG} = \frac{1}{4\pi\varepsilon_0}\frac{q}{r_{minus}^2} \qquad E_{plusMAG} = \frac{1}{4\pi\varepsilon_0}\frac{q}{r_{plus}^2}$$

Step 5:   Create the vectors $\mathbf{E}_-$ and $\mathbf{E}_+$ by multiplying the unit vectors by the magnitudes.

Step 6:   Create the vector **E** by adding the vectors $\mathbf{E}_-$ and $\mathbf{E}_+$.

Step 7:   Calculate E, the magnitude (length) of **E**.

Step 8:   Plot E as a function of x.

A program in a script file that solves the problem is:

```
q=12e-9;
epsilon0=8.8541878e-12;                          Create a column vector x.
x=[-0.05:0.001:0.05]';
rminusS=(0.02-x).^2+0.02^2; rminus=sqrt(rminusS); Step 2, each variable
                                                  is a column vector.
rplusS=(x+0.02).^2+0.02^2; rplus=sqrt(rplusS);   Steps 3 & 4, each variable
EminusUV=[((0.02-x)./rminus), (-0.02./rminus)];  is a two column matrix.
                                                  Each row is the vector for
EplusUV=[((x+0.02)./rplus), (0.02./rplus)];       the corresponding x.
```

```
EminusMAG=(q/(4*pi*epsilon0))./rminusS;
EplusMAG=(q/(4*pi*epsilon0))./rplusS;
Eminus=[EminusMAG.*EminusUV(:,1), EminusMAG.*EminusUV(:,2)];
Eplus=[EplusMAG.*EplusUV(:,1), EplusMAG.*EplusUV(:,2)];
E=Eminus+Eplus;
EMAG=sqrt(E(:,1).^2+E(:,2).^2);
plot(x,EMAG,'k','linewidth',1)
xlabel('Position along the x-axis (m)','FontSize',12)
ylabel('Magnitude of the electric field (N/C)','FontSize',12)
title('ELECTRIC FIELD DUE TO AN ELECTRIC DIPOLE','FontSize',12)
```

Step 6.

Step 7.

Step 5.

When this script file is executed in the Command Window the following figure is created in the Figure Window:

5.13 PROBLEMS

1. Make two separate plots of the function $f(x) = 0.01x^4 - 0.45x^2 + 0.5x - 2$; one plot for $-4 \le x \le 4$, and one for $-8 \le x \le 8$.

2. Plot the function $f(t) = \dfrac{5}{1 + e^{5.5 - 1.5x}} - \dfrac{x^2}{20}$ for $-10 \le x \le 10$.

3. Use the `fplot` command to plot the function:
   $$f(x) = \frac{40}{1 + (x-4)^2} + 5\sin\left(\frac{20x}{\pi}\right) \text{ in the domain } 0 \le x \le 10.$$

4. Plot the function $f(x) = \dfrac{x^2 - 4x - 5}{x - 2}$ for $-4 \le x \le 8$. Notice that the function has a vertical asymptote at $x = 2$. Plot the function by creating two vectors for the domain of x, the first vector (call it $x1$) with elements from -4 to 1.7, and

the second vector (call it *x2*) with elements from 2.3 to 8. For each *x* vector create a *y* vector (call them *y1* and *y2*) with the corresponding values of *y* according to the function. To plot the function make two curves in the same plot (*y1* vs. *x1*, and *y2* vs. *x2*).

5. Plot the function $f(x) = \dfrac{4x - 30}{x^2 - 3x - 10}$ for $-10 \le x \le 10$. Notice that the function has two vertical asymptotes. Plot the function by dividing the domain of *x* into three parts: one from -10 to near the left asymptote, one between the two asymptotes, and one from near the right asymptote to 10. Set the range of the *y*-axis from -20 to 20.

6. Plot the function $f(x) = 3x\cos^2 x - 2x$ and its derivative, both on the same plot, for $-2\pi \le x \le 2\pi$. Plot the function with a solid line, and the derivative with a dashed line. Add a legend and label the axes.

7. An electrical circuit that includes a voltage source v_S with an internal resistance r_S and a load resistance R_L is shown in the figure. The power *P* dissipated in the load is given by:

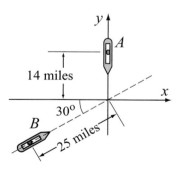

$$P = \frac{v_S^2 R_L}{(R_L + r_S)^2}$$

Plot the power *P* as a function of R_L for $1 \le R_L \le 10\,\Omega$, given that $v_S = 12$ V, and $r_S = 2.5\,\Omega$.

8. Ship *A* travels south at a speed of 6 miles/hour, and a ship *B* travels 30° north to the east at a speed of 14 miles/hour. At 7 AM the ships are positioned as shown in the figure. Plot the distance between the ships as a function of time for the next 4 hours. The horizontal axis should show the actual time of day starting at 7 AM, while the vertical axis shows the distance. Label the axes. If visibility is 8 miles, estimate from the graph the time when people from the two ships can see each other.

9. The Gateway Arch in St. Louis is shaped according to the equation:

$$y = 693.8 - 68.8\cosh\left(\frac{x}{99.7}\right) \text{ ft.}$$

Make a plot of the arch.

10. The position as a function of time $[x(t), y(t)]$ of a squirrel running on a grass field is given by:

$$x(t) = -0.28t^2 + 6.5t + 61 \text{ m and}$$

$$y(t) = 0.18t^2 - 8.5t + 65 \text{ m}$$

a) Plot the trajectory (position) of the squirrel for $0 \le t \le 30$ s.

b) Plot (a second plot) the length of the position vector $r(t) = x(t)i + y(t)j$ of the squirrel for $0 \le t \le 30$ s.

c) Plot (a third plot) the angle $\theta(t)$ of the position vector for $0 \le t \le 30$ s.

11. In astronomy, the relationship between the relative luminosity L/L_{Sun} (brightness relative to the sun), the relative radius R/R_{Sun}, and the relative temperature T/T_{Sun} of a star is modeled by:

$$\frac{L}{L_{Sun}} = \left(\frac{R}{R_{Sun}}\right)^2 \left(\frac{T}{T_{Sun}}\right)^4$$

The HR (Hertzsprung-Russell) diagram is a plot of L/L_{Sun} versus the temperature. The following data is given:

	Sun	Spica	Regulus	Alioth	Barnard's Star	Epsilon Indi	Beta Crucis
Temp (K)	5840	22400	13260	9400	3130	4280	28200
L/L_{sun}	1	13400	150	108	0.0004	0.15	34000
R/R_{sun}	1	7.8	3.5	3.7	0.18	0.76	8

To compare the data with the model, use MATLAB to plot a HR diagram. The diagram should have two sets of points. One uses the values of L/L_{Sun} from the Table (use asterisk markers), and the other uses values of L/L_{Sun} that are calculated by the equation by using R/R_{Sun} from the Table (use circle markers). In the HR diagram both axes are logarithmic. In addition, the values of the temperature on the horizontal axis are decreasing from left to right. This is done with the command: set(gca,'XDir','reverse'). Label the axes and use a legend.

12. The position x as a function of time of a particle that moves along a straight line is given by:

$$x(t) = -0.1t^4 + 0.8t^3 + 10t - 70 \text{ m}$$

The velocity $v(t)$ of the particle is determined by the derivative of $x(t)$ with respect to t, and the acceleration $a(t)$ is determined by the derivative of $v(t)$ with respect to t.

Derive the expressions for the velocity and acceleration of the particle, and make plots of the position, velocity, and acceleration as a function of time for $0 \le t \le 8$ s. Use the `subplot` command to make the three plots on the same page with the plot of the position on the top, the velocity in the middle, and the acceleration at the bottom. Label the axes appropriately with the correct units.

13. In a typical tension test a dog-bone-shaped specimen is pulled in a machine. During the test, the force F needed to pull the specimen and the length L of a gauge section are measured. This data is used for plotting a stress-strain diagram of the material. Two definitions, engineering and true, exist for stress and strain. The engineering stress σ_e and strain ε_e are defined by:

$$\sigma_e = \frac{F}{A_0} \text{ and } \varepsilon_e = \frac{L - L_0}{L_0}, \text{ where } L_0 \text{ and } A_0 \text{ are the initial gauge length and the}$$

initial cross-sectional area of the specimen, respectively. The true stress σ_t and strain ε_t are defined by: $\sigma_t = \dfrac{F}{A_0} \dfrac{L}{L_0}$ and $\varepsilon_t = \ln \dfrac{L}{L_0}$.

    The following are measurements of force and gauge length from a tension test with an aluminum specimen. The specimen has a round cross section with radius of 6.4 mm (before the test). The initial gauge length is $L_0 = 25$ mm. Use the data to calculate and plot the engineering and true stress-strain curves, both on the same plot of the material. Label the axes and label the curves.
    Units: When the force is measured in Newtons (N), and the area is calculated in m², the units of the stress are Pascals (Pa).

F (N)	0	13345	26689	40479	42703	43592	44482	44927
L (mm)	25	25.037	25.073	25.113	25.122	25.125	25.132	25.144
F (N)	45372	46276	47908	49035	50265	53213	56161	
L (mm)	25.164	25.208	25.409	25.646	26.084	27.398	29.150	

14. The area of the aortic valve, A_V in cm², can be estimated by the equation (Hakki Formula):

$$A_V = \frac{Q}{\sqrt{PG}}$$

where Q is the cardiac output in L/min, and PG is the difference between the left ventricular systolic pressure and the aortic systolic pressure (in mmHg). Make one plot with two curves of A_V versus PG, for $2 \le PG \le 60$ mmHg. One curve for $Q = 4$ L/min and the other for $Q = 5$ L/min. Label the axes and use a legend.

15. A resistor, $R = 4\ \Omega$, and an inductor, $L = 1.3$ H, are connected in a circuit to a voltage source as shown in Figure (a) $(RL$ circuit$)$. When the voltage source

(a)

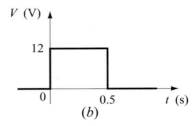

(b)

applies a rectangular voltage pulse with an amplitude of $V = 12$ V and a duration of 0.5 s, as shown in Figure (b), the current $i(t)$ in the circuit as a function of time is given by:

$$i(t) = \frac{V}{R}(1 - e^{(-Rt)/L}) \quad \text{for } 0 \le t \le 0.5 \text{ s}$$

$$i(t) = e^{-(Rt)/L}\frac{V}{R}(e^{(0.5R)/L} - 1) \quad \text{for } 0.5 \le t \text{ s}$$

Make a plot of the current as a function of time for $0 \le t \le 2$ s.

16. The dynamic storage modulus G' and loss modulus G'' are measures of a material mechanical response to harmonic loading. For many biological materials these moduli can be described by Fung's model:

$$G'(\omega) = G_\infty\left\{1 + \frac{c}{2}\ln\left[\frac{1 + (\omega\tau_2)^2}{1 + (\omega\tau_1)^2}\right]\right\} \quad \text{and} \quad G''(\omega) = cG_\infty[\tan^{-1}(\omega\tau_2) - \tan^{-1}(\omega\tau_1)]$$

where ω is the frequency of the harmonic loading, and G_∞, c, τ_1, and τ_2 are material constants. Plot G' and G'' versus ω (two separate plots on the same page) for $G_\infty = 5$ ksi, $c = 0.05$, $\tau_1 = 0.05$s, and $\tau_2 = 500$ s. Let ω vary between 0.0001 and 1000 s^{-1}. Use log scale for the ω axis.

17. The vibrations of the body of a helicopter due to the periodic force applied by the rotation of the rotor can be modeled by a frictionless spring-mass-damper system subjected to an external periodic force. The position $x(t)$ of the mass is given by the equation:

$$x(t) = \frac{2f_0}{\omega_n^2 - \omega^2}\sin\left(\frac{\omega_n - \omega}{2}t\right)\sin\left(\frac{\omega_n - \omega}{2}t\right)$$

where $F(t) = F_0\sin\omega t$, and $f_0 = F_0/m$, ω is frequency of the applied force, and ω_n is the natural frequency of the helicopter. When the value of ω is close to the value of ω_n the vibration consists of fast oscillation with slowly changing amplitude called beat. Use $F_0/m = 12\,\text{N/kg}$, $\omega_n = 10\,\text{rad/s}$, and $\omega = 12\,\text{rad/s}$ to plot $x(t)$ as a function of t for $0 \le t \le 10$ s.

18. The ideal gas equation states that $\dfrac{PV}{RT} = n$, where P is the pressure, V is the volume, T is the temperature, $R = 0.08206\,\text{(L atm)/(mole K)}$ is the gas constant, and n is the number of moles. For one mole ($n = 1$) the quantity $\dfrac{PV}{RT}$ is a constant equal to 1 at all pressures. Real gases, especially at high pressures, deviate from this behavior. Their response can be modeled with the van der Waals equation:

$$P = \frac{nRT}{V - nb} - \frac{n^2 a}{V^2}$$

where a and b are material constants. Consider 1 mole ($n = 1$) of nitrogen gas at $T = 300\,\text{K}$. (For nitrogen gas $a = 1.39\,\text{L}^2\text{atm/mole}^2$, and $b = 0.0391$ L/mole.) Use the van der Waals equation to calculate P as a function of V for $0.08 \le V \le 6\,\text{L}$, using increments of 0.02L. At each value of V calculate the value of $\dfrac{PV}{RT}$ and make a plot of $\dfrac{PV}{RT}$ versus P. Does the response of nitrogen agrees with the ideal gas equation?

19. A simply supported beam that is sub-jected to a constant distributed load w over half of its length is shown in the figure. The deflection y, as a function of x, is given by the equations:

$$y = \frac{-wx}{24LEI}\left(Lx^3 - \frac{16}{9}L^2x^2 + \frac{64}{81}L^4\right) \text{ for } 0 \le x \le \frac{2}{3}L$$

$$y = \frac{-wL}{54EI}\left(2x^3 - 6Lx^2 + \frac{40}{9}L^2x - \frac{4}{9}L^3\right) \text{ for } \frac{2}{3}L \le x \le L$$

where E is the elastic modulus, I is the moment of inertia, and L is the length of the beam. For the beam shown in the figure $L = 20$ m, $E = 200 \times 10^9$ Pa (steel), $I = 348 \times 10^{-6}$ m⁴, and $w = 5 \times 10^3$ N/m. Make a plot of the deflection of the beam y as a function of x.

Chapter 6
User-Defined Functions and Function Files

A simple function in mathematics, $f(x)$, associates a unique number to each value of x. The function can be expressed in the form $y = f(x)$, where $f(x)$ is usually a mathematical expression in terms of x. A value of y (output) is obtained when a value of x (input) is substituted in the expression. Many functions are programed inside MATLAB as built-in functions, and can be used in mathematical expressions simply by typing their name with an argument (see Section 1.5); examples are `sin(x)`, `cos(x)`, `sqrt(x)`, and `exp(x)`. Frequently, in computer programs, there is a need to calculate the value of functions that are not built-in. When a function expression is simple and needs to be calculated only once, it can be typed as part of the program. However, when a function needs to be evaluated many times for different values of arguments it is convenient to create a "user-defined" function. Once a user-defined function is created (saved) it can be used just like the built-in functions.

A user-defined function is a MATLAB program that is created by the user, saved as a function file, and then can be used like a built-in function. The function can be a simple single mathematical expression or a complicated and involved series of calculations. In many cases it is actually a subprogram within a computer program. The main feature of a function file is that it has an input and an output. This means that the calculations inside the function file are carried out using the input data, and the results of the calculations are transferred out of the function file by the output. The input and the output can be one or several variables, and each can be a scalar, vector, or an array of any size. Schematically, a function file can be illustrated by:

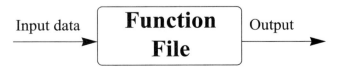

A very simple example of a user-defined function is a function that calculates the maximum height that a ball reaches when thrown up with a certain velocity. For a velocity v_0, the maximum height h_{max} is given by: $h_{max} = \dfrac{v_0^2}{2g}$, where g is the gravitational acceleration. In a function form this can be written as: $h_{max}(v_0) = \dfrac{v_0^2}{2g}$. In this case the input to the function is the velocity (a number), and the output is the maximum height (a number). For example, in SI units (g = 9.81 m/s^2) if the input is 15 m/s, the output is 11.47 m.

$$\xrightarrow{\text{15 m/s}} \boxed{\text{Function File}} \xrightarrow{\text{11.47 m}}$$

In addition to being used as a math function, user-defined functions can be used as subprograms in large programs. In this way large computer programs can be made up of smaller "building blocks" that can be tested independently. Function files are similar to subroutines in Basic and Fortran, procedures in Pascal, and functions in C.

The fundamentals of user-defined functions are explained in Sections 6.1 through 6.7. In addition to user-defined functions that are saved in separate function files and are called for use in a computer program, MATLAB provides an option to define and use a user-defined math function within a computer program (not in a separate file). This can be done by using anonymous and/or inline functions, which are presented in Section 6.8. There are built-in and user-defined functions that have to be supplied with other functions when they are called. These functions, which are called function functions in MATLAB, are introduced in Section 6.9. The last two sections cover subfunctions and nested functions. Both are methods for incorporating two, or more, user-defined functions in a single function file.

6.1 CREATING A FUNCTION FILE

Function files are created and edited, like script files, in the Editor/Debugger Window. This window is opened from the Command Window. In the **File** menu, select **New**, and then select **M-file**. Once the Editor/Debugger Window opens it looks like that shown in Figure 6-1, and the commands of the file can be typed line after line. The first line in a function file must be the function definition line, as discussed in the next section.

6.2 STRUCTURE OF A FUNCTION FILE

The structure of a typical function file is shown in Figure 6-2. This particular function calculates the monthly payment and the total payment of a loan. The inputs to the function are the amount of the loan, the annual interest rate, and the duration of the loan (number of years of the loan). The output from the function is

Figure 6-1: The Editor/Debugger Window.

the monthly payment and the total payment.

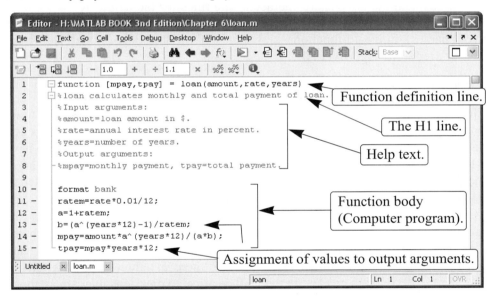

Figure 6-2: Structure of a typical function file.

The various parts of the function file are described in detail in the following sections.

6.2.1 Function Definition Line

The first executable line in a function file **must** be the function definition line. Otherwise the file is considered a script file. The function definition line:

• Defines the file as a function file.

• Defines the name of the function.

• Defines the number and order of the input and output arguments.

The form of the function definition line is:

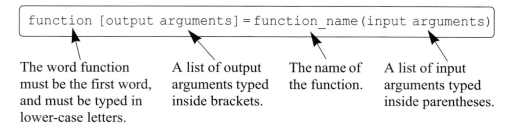

The word function, typed in lower case letters, must be the first word in the function definition line. On the screen the word function appears in blue. The function name is typed following the equal sign. The name can be made up of letters, digits, and the underscore character. The rules for the name are the same as the rules for naming variables as described in Section 1.6.2. It is good practice to avoid names of built-in functions, and names of variables already defined by the user or predefined by MATLAB.

6.2.2 Input and Output Arguments

The input and output arguments are used to transfer data into and out of the function. The input arguments are listed inside parentheses following the function name. Usually, there is at least one input argument, although it is possible to have a function that has no input arguments. If there are more than one, the input arguments are separated with commas. The computer code that performs the calculations within the function file is written in terms of the input arguments and assumes that the arguments have assigned numerical values. This means that the mathematical expressions in the function file must be written according to the dimensions of the arguments, since the arguments can be scalars, vectors, and arrays. In the example shown in Figure 6-2 there are three input arguments, (amount, rate, years) and in the mathematical expressions they are assumed to be scalars. The actual values of the input arguments are assigned when the function is used (called). Similarly, if the input arguments are vectors or arrays, the mathematical expressions in the function body must be written to follow linear algebra or element-by-element calculations.

The output arguments, which are listed inside brackets on the left side of the assignment operator in the function definition line, transfer the output from the function file. Function files can have none, one, or several output arguments. If there are more than one, the output arguments are separated with commas. If there is only one output argument it can be typed without brackets. **In order for the function file to work, the output arguments must be assigned values in the computer program that is in the function body.** In the example in Figure 6-2 there are two output arguments, [mpay, tpay]. When a function does not have an output argument, the assignment operator in the function definition line can be omitted. A function without an output argument can, for example, generate a plot or write data to a file.

It is also possible to transfer strings into a function file. This is done by typing the string as part of the input variables (text enclosed in single quotes). Strings can be used to transfer names of other functions into the function file.

Usually, all the input to, and the output from, a function file are transferred through the input and output arguments. In addition, however, all the input and output features of script files are valid and can be used in function files. This means that any variable that is assigned a value in the code of the function file will be displayed on the screen unless a semicolon is typed at the end of the command. In addition, the `input` command can be used to input data interactively, and the `disp`, `fprintf`, and `plot` commands can be used to display information on the screen, save to a file, or plot figures just as in a script file. The following are examples of function definition lines with different combinations of input and output arguments.

Function definition line	**Comments**
function[mpay,tpay] = loan(amount,rate,years)	Three input arguments, two output arguments.
function [A] = RectArea(a,b)	Two input arguments, one output argument.
function A = RectArea(a,b)	Same as above, one output argument can be typed without the brackets.
function [V, S] = SphereVolArea(r)	One input variable, two output variables.
function trajectory(v,h,g)	Three input arguments, no output arguments.

6.2.3 The H1 Line and Help Text Lines

The H1 line and help text lines are comment lines (lines that begin with the percent % sign) following the function definition line. They are optional, but frequently used to provide information about the function. The H1 line is the first comment line and usually contains the name and a short definition of the function. When a user types (in the Command Window) `lookfor a_word`, MATLAB searches for `a_word` in the H1 lines of all the functions, and if a match is found, the H1 line that contains the match is displayed.

The help text lines are comment lines that follow the H1 line. These lines contain an explanation of the function and any instructions related to the input and output arguments. The comment lines that are typed between the function definition line and the first non-comment line (the H1 line and the help text) are displayed when the user types `help function_name` in the Command Window. This is true for MATLAB built-in functions as well as the user-defined functions. For example, for the function `loan` in Figure 6-2, if `help loan` is typed in the Command Window (make sure the current directory or the search path includes the directory where the file is saved), the display on the screen is:

```
>> help loan
loan calculates monthly and total payment of loan.
Input arguments:
amount=loan amount in $.
rate=annual interest rate in percent.
years=number of years.
Output arguments:
mpay=monthly payment, tpay=total payment.
```

A function file can include additional comment lines in the function body. These lines are ignored by the `help` command.

6.2.4 Function Body

The function body contains the computer program (code) that actually performs the computations. The code can use all MATLAB programming features. This includes calculations, assignments, any built-in or user-defined functions, flow control (conditional statements and loops) as explained in Chapter 7, comments, blank lines, and interactive input and output.

6.3 LOCAL AND GLOBAL VARIABLES

All the variables in a function file are local (the input and output arguments and any variables that are assigned values within the function file). This means that the variables are defined and recognized only inside the function file. When a function file is executed, MATLAB uses an area of memory that is separate from the workspace (the memory space of the Command Window and the script files). In a function file the input variables are assigned values each time the function is called. These variables are then used in the calculations within the function file. When the function file finishes its execution the value of the output arguments are transferred to the variables that were used when the function was called. All of this means that a function file can have variables with the same name as variables in the Command Window or in script files. The function file does not recognize variables with the same name that have been assigned values outside the function. The assignment of values to these variables in the function file will not change their assignment elsewhere.

Each function file has its own local variables which are not shared with other functions or with the workspace of the Command Window and the script files. It is possible, however, to make a variable common (recognized) in several different function files, and perhaps in the workspace too. This is done by declaring the variable global with the `global` command that has the form:

```
global variable_name
```

Several variables can be declared global by listing them, separated with spaces, in the global command. For example:

```
global GRAVITY_CONST FrictionCoefficient
```

- The variable has to be declared global in every function file that the user wants it to be recognized in. The variable is then common only to these files.

- The `global` command must appear before the variable is used. It is recommended to enter the `global` command at the top of the file.

- The `global` command has to be entered in the Command Window, or/and in a script file, for the variable to be recognized in the workspace.

- The variable value can be assigned, or reassigned, a value in any of the locations it is declared common.

- It is recommended to use long descriptive names (or use all capital letters) for global variables in order to distinguish between them and regular variables.

6.4 SAVING A FUNCTION FILE

A function file must be saved before it can be used. This is done, as with a script file, by choosing **Save as ...** from the **File** menu, selecting a location (many students save to an A drive or a zip drive), and entering the file name. It is highly recommended that the file is saved with a name that is identical to the function name in the function definition line. In this way the function is called (used) by using the function name. (If a function file is saved with a different name, the name it is saved under must be used when the function is called.) Function files are saved with the extension .m. Examples:

Function definition line	**File name**
function[mpay,tpay] = loan(amount,rate,years)	loan.m
function [A] = RectArea(a,b)	RectArea.m
function [V, S] = SphereVolArea(r)	SphereVolArea.m
function trajectory(v,h,g)	trajectory.m

6.5 *USING A USER-DEFINED FUNCTION*

A user-defined function is used in the same way as a built-in function. The function can be called from the Command Window, from a script file, or from another function. To use the function file, the directory where it was saved must either be in the current directory or be in the search path (see Sections 4.3.1 and 4.3.2).

A function can be used by assigning its output to a variable (or variables), as a part of a mathematical expression, as an argument in another function, or just by typing its name in the Command Window or in a script file. In all cases the user must know exactly what the input and output arguments are. An input argument can be a number, a computable expression, or it can be a variable that has an assigned value. The arguments are assigned according to their position in the input and output argument lists in the function definition line.

Two of the ways that a function can be used are illustrated below with the user-defined `loan` function in Figure 6-2 that calculates the monthly and total payment (two output arguments) of a loan. The input arguments are the loan amount, annual interest rate, and the length (number of years) of the loan. In the first illustration the `loan` function is used with numbers as input arguments:

```
>> [month total]=loan(25000,7.5,4)
```

> First argument is loan amount, second is interest rate, and third is number of years.

```
month =
        600.72
total =
      28834.47
```

In the second illustration the `loan` function is used with two preassigned variables and a number as the input arguments:

```
>> a=70000;  b=6.5;
```

> Define variables a and b.

```
>> [x y]=loan(a,b,30)
```

> Use a, b, and the number 30 for input arguments and x (monthly pay) and y (total pay) for output arguments.

```
x =
        440.06
y =
      158423.02
```

6.6 EXAMPLES OF SIMPLE USER-DEFINED FUNCTIONS

Sample Problem 6-1:   User-defined function of a math function

Write a function file (name it chp6one) for the function $f(x) = \dfrac{x^4 \sqrt{3x+5}}{(x^2+1)^2}$. The input to the function is x and the output is $f(x)$. Write the function such that x can be a vector. Use the function to calculate:

a) $f(x)$ for $x = 6$.

b) $f(x)$ for $x = 1, 3, 5, 7, 9,$ and 11.

Solution

The function file for the function $f(x)$ is:

```
function y=chp6one(x)                        Function definition line.
y=(x.^4.*sqrt(3*x+5))./(x.^2+1).^2;   Assignment to output argument.
```

Note that the mathematical expression in the function file is written for element-by-element calculations. In this way if x is a vector y will also be a vector. The function is saved and then the search path is modified to include the directory where the file was saved. As shown below the function is used in the Command Window.

a)   Calculating the function for $x = 6$ can be done by typing chp6one(6) in the Command Window, or by assigning the value of the function to a new variable:

```
>> chp6one(6)
ans =
    4.5401
>> F=chp6one(6)
F =
    4.5401
```

b)   To calculate the function for several values of x, a vector with the values of x is first created, and then used for the argument of the function.

```
>> x=1:2:11
x =
     1     3     5     7     9    11
>> chp6one(x)
ans =
    0.7071    3.0307    4.1347    4.8971    5.5197    6.0638
```

Another way is to type the vector x directly in the argument of the function.

```
>> H=chp6one([1:2:11])
H =
    0.7071    3.0307    4.1347    4.8971    5.5197    6.0638
```

Sample Problem 6-2: Converting temperature units

Write a user-defined function (name it FtoC) that converts temperature in degrees
F to temperature in degrees C. Use the function to solve the following problem.
The change in the length of an object, ΔL, due to a change in the temperature, ΔT,
is given by: $\Delta L = \alpha L \Delta T$, where α is the coefficient of thermal expansion. Deter-
mine the change in the area of a rectangular (4.5 m by 2.25 m) aluminum
($\alpha = 23 \cdot 10^{-6}$ $1/^{\circ}$C) plate if the temperature changes from 40°F to 92°F.

Solution

A user-defined function that converts degrees F to degrees C is:

```
function C=FtoC(F)                          Function definition line.
%FtoC converts degrees F to degrees C
C=5*(F-32)./9;                       Assignment to output argument.
```

A script file (named Chapter6Example2) that calculates the change of the area of
the plate due to the temperature is:

```
a1=4.5; b1=2.25; T1=40; T2=92; alpha=23e-6;
deltaT=FtoC(T2)-FtoC(T1);     Using the FtoC function to calculate the
                              temperature difference in degrees C.
a2=a1+alpha*a1*deltaT;             Calculating the new length.
b2=b1+alpha*b1*deltaT;             Calculating the new width.
AreaChange=a2*b2-a1*b1;         Calculating the change in the area.
fprintf('The change in the area is %6.5f meters
square.',AreaChange)
```

Executing the script file in the Command Window gives the solution:

```
>> Chapter6Example2
The change in the area is 0.01346 meters square.
```

6.7 *COMPARISON BETWEEN SCRIPT FILES AND FUNCTION FILES*

Students who are studying MATLAB for the first time sometimes have difficulty understanding exactly the differences between script and function files since, for many of the problems that they are asked to solve by using MATLAB, either type of file can be used. The similarities and differences between script and function files are summarized below.

- Both script and function files are saved with the extension .m (that is why they are sometimes called M-files).

- The first line in a function file is the function definition line.

- The variables in a function file are local. The variables in a script file are recognized in the Command Window.

- Script files can use variables that have been defined in the workspace.

- Script files contain a sequence of MATLAB commands (statements).

- Function files can accept data through input arguments and can return data through output arguments.

- When a function file is saved, the name of the file should be the same as the name of the function.

6.8 *ANONYMOUS AND INLINE FUNCTIONS*

User-defined functions written in function files can be used for simple mathematical functions, for large and complicated math functions that require extensive programming, and as subprograms in large computer programs. In cases when the value of a relatively simple mathematical expression has to be determined many times within a program, MATLAB provides the option of using anonymous functions. An anonymous function is a user-defined function that is defined and written within the computer code (not in a separate function file) and is then used in the code. Anonymous functions can be defined in any part of MATLAB (in the Command Window, in script files, and inside regular user-defined functions).

Anonymous functions have been introduced in MATLAB 7. They replace inline functions that have been used for the same purpose in previous versions of MATLAB. Both anonymous and inline functions can be used in MATLAB 7.5 (R2007b). Anonymous functions, however, have several advantages over inline functions, and it can be expected that inline functions will gradually be phased out. Anonymous functions are covered in detail in Sections 6.8.1, and inline functions are described in the section that follows.

6.8.1 Anonymous Functions

An anonymous function is a simple (one-line) user-defined function that is defined without creating a separate function file (M-file). Anonymous functions can be constructed in the Command Window, within a script file, or inside a regular user-defined function.

An anonymous function is created by typing the following command:

```
name = @ (arglist) expr
```

The name of the anony- The @ A list of input argu- Mathematical
mous function. symbol. ments (independent expression.
variables).

A simple example is: `cube = @ (x) x^3`, which calculates the cubic power of the input argument.

- The command creates the anonymous function, and assigns a handle for the function to the variable name that is on the left-hand side of the = sign. (Function handles provide means for using the function, and passing it to other functions, see Section 6.9.1.)

- The `expr` consists of a single valid mathematical MATLAB expression.

- The mathematical expression can have one or several independent variables. The independent variable(s) is (are) listed in the `(arglist)`. If there are more than one, the independent variables are separated with a comma.
  An example of an anonymous function that has two independent variables is: `circle = @ (x,y) 16*x^2+9*y^2`

- The mathematical expression can include any built-in or user-defined functions.

- The expression must be written according to the dimensions of the arguments (element-by-element or linear algebra calculations).

- The expression can include predefined variables that are already defined when the anonymous function is defined. For example, if three variables a, b, and c are defined (they have assigned numerical values), then they can be used in the expression of the anonymous function: `parabola = @ (x) a*x^2+b*x+c`. **Important note:** MATLAB captures the values of the predefined variables when the anonymous function is defined. This means that if subsequently new values are assigned to the predefined variables, the anonymous function is not changed. The anonymous function has to be redefined in order for the new values of the predefined variables to be used in the expression.

Using an anonymous function:

- Once an anonymous function is defined, it can be used by typing its name and a value for the argument (or arguments) in parentheses (see examples that follow).

- Anonymous functions can also be used as arguments in other functions (see Section 6.9.1).

Example of an anonymous function with one independent variable:

The function: $f(x) = \dfrac{e^{x^2}}{\sqrt{x^2+5}}$ can be defined (in the Command Window) as an inline function for x as a scalar by:

```
>> FA = @ (x) exp(x^2)/sqrt(x^2+5)
FA =
    @(x)exp(x^2)/sqrt(x^2+5)
```

If a semicolon is not typed at the end, MATLAB responds by displaying the function. The function can then be used for different values of x, as shown below.

```
>> FA(2)
ans =
   18.1994
>> z = FA(3)
z =
   2.1656e+003
```

If x is expected to be an array, and the function calculated for each element, then the function must be modified for element-by-element calculations.:

```
>> FA = @ (x) exp(x.^2)./sqrt(x.^2+5)
FA =
    @(x)exp(x.^2)./sqrt(x.^2+5)
>> FA([1 0.5 2])            Using a vector as input argument.
ans =
    1.1097    0.5604   18.1994
```

Example of an anonymous function with several independent variables:

The function $f(x, y) = 2x^2 - 4xy + y^2$ can be defined as an anonymous function by:

```
>> HA = @ (x,y) 2*x^2 - 4*x*y + y^2
HA =
    @(x,y)2*x^2-4*x*y+y^2
```

Then, the anonymous function can be used for different values of *x* and *y*. For example, typing HA(2,3) gives:

```
>> HA(2,3)
ans =
    -7
```

Another example of using an anonymous function with several arguments is shown next in Sample Problem 6-3.

Sample Problem 6-3:   Distance between points in polar coordinates

Write an anonymous function that calculates the distance between two points in a plane when the position of the points is given in polar coordinates. Use the anonymous function to calculate the distance between point *A* (2, π/6) and point *B* (5, 3π/4).

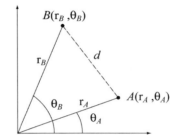

Solution

The distance between two points in polar coordinates can be calculated by using the law of cosines:

$$d = \sqrt{r_A^2 + r_B^2 - 2r_A r_B \cos(\theta_A - \theta_B)}$$

The formula for the distance is first entered as an anonymous function with four input arguments $(r_A, \theta_A, r_B, \theta_B)$. Then the function is used for calculating the distance between points *A* and *B*.

```
>> d= @ (rA,thetA,rB,thetB) sqrt(rA^2+rB^2-2*rA*rB*cos(thetB-thetA))
```
List of input arguments.
```
d =
    @ (rA,thetA,rB,thetB) sqrt(rA^2+rB^2-2*rA*rB*cos(thetB-
thetA))
>> DistAtoB = d(2,pi/6,5,3*pi/4)
DistAtoB =
    5.8461
```
The arguments are typed in the order defined in the function.

6.8.2 *Inline Functions*

Similar to anonymous function, an inline function is a simple user-defined function that is defined without creating a separate function file (M-file). As already mentioned, anonymous functions replace inline functions used in earlier versions of MATLAB. Inline functions are created with the `inline` command according to the following format:

```
name = inline('math expression typed as a string')
```

A simple example is: `cube = inline('x^3')`, which calculates the cubic power of the input argument.

- The mathematical expression can have one or several independent variables.

- Any letter except i and j can be used for the independent variables in the expression.

- The mathematical expression can include any built-in or user-defined functions.

- The expression must be written according to the dimension of the argument (element-by-element or linear algebra calculations).

- The expression <u>cannot</u> include preassigned variables.

- Once the function is defined it can be used by typing its name and a value for the argument (or arguments) in parentheses (see example below).

- The `inline` function can also be used as an argument in other functions.

For example, the function: $f(x) = \dfrac{e^{x^2}}{\sqrt{x^2 + 5}}$ can be defined as an inline function for x by:

```
>> FA=inline('exp(x.^2)./sqrt(x.^2+5)')
FA =
     Inline function:
     FA(x) = exp(x.^2)./sqrt(x.^2+5)
>> FA(2)
ans =
   18.1994
>> FA([1 0.5 2])
ans =
    1.1097    0.5604   18.1994
```

Expression written with element-by-element operations.

Using a scalar as the argument.

Using a vector as the argument.

An inline function that has two or more independent variables can be written by using the following format:

```
name = inline('mathematical expression','arg1',
                                  'arg2','arg3')
```

In the format shown here the order of the arguments to be used when calling the function is defined. If the independent variables are not listed in the command, MATLAB arranges the arguments in alphabetical order. For example, the $f(x, y) = 2x^2 - 4xy + y^2$ can be defined as an inline function by:

```
>> HA=inline('2*x^2-4*x*y+y^2')

HA =

     Inline function:
     HA(x,y) = 2*x^2-4*x*y+y^2
```

Once defined, the function can be used with any values of x and y. For example, HA(2,3) gives:

```
>> HA(2,3)
ans =
    -7
```

6.9 FUNCTION FUNCTIONS

There are many situations where a function (Function A) works on (uses) another function (Function B). This means that when Function A is executed it has to be provided with Function B. A function that accepts another function is called in MATLAB a function function. For example, MATLAB has a built-in function called fzero (Function A) that finds the zero of a math function $f(x)$ (Function B), i.e., the value of x where $f(x) = 0$. The program in the function fzero is written such that it can find the zero of any $f(x)$. When fzero is called, the specific function to be solved is passed into fzero, which finds the zero of the $f(x)$. (The function fzero is described in detail in Chapter 10.)

A function function, which accepts another function (imported function), includes in its input arguments a name that represents the imported function. The imported function name is used for the operations in the program (code) of the function function. When the function function is used (called), the specific function that is imported is listed in its input argument. In this way different functions can be imported (passed) into the function function. There are two methods for listing the name of an imported function in the argument list of a function function. One is by using a function handle (Section 6.9.1), and the other is by typing the name of the function that is being passed in as a string expression (Section 6.9.2). The method that is used affects the way that the operations in the function

function are written (this is explained in more detail in the next two sections). Using function handles is easier and more efficient, and should be the preferred method to use.

6.9.1 Using Function Handles for Passing a Function into a Function Function

Function handles are used for passing (importing) user-defined functions, built-in functions, and anonymous functions into function functions that can accept them. This section first explains what a function handle is, then shows how to write a user-defined function function that accepts function handles, and finally shows how to use function handles for passing functions into function functions.

Function handle:

A function handle is a MATLAB value that is associated to a function. It is a MATLAB data type, and can be passed as an argument into another function. Once passed, the function handle provides means for calling (using) the function it is associated with. Function handles can be used with any kind of MATLAB function. This includes built-in functions, user-defined functions (written in function files), and anonymous functions.

- For built-in and user-defined functions, a function handle is created by typing the symbol @ in front of the function name. For example, @cos is the function handle of the built-in function cos, and @FtoC is the function handle of the user-defined function FtoC that was written in Sample Problem 6-2.

- The function handle can also be assigned to a variable name. For example, cosHandle=@cos assigns the handle @cos to cosHandle. Then, the name cosHandle can be used for passing the handle.

- For anonymous functions (see Section 6.8.1), their name is already a function handle.

Writing a function function that accepts a function handle as an input argument:

As already mentioned, the input arguments of a function function (that accepts another function) includes a name (dummy function name) that represents the imported function. This dummy function (including a list of input arguments enclosed in parentheses) is used for the operations in the program inside the function function.

- The function that is actually being imported must be consistent with the way that the dummy function is being used in the program. This means that both must have the same number and type of input and output arguments.

The following is an example of a user-defined function function, named funplot, that makes a plot of a function (any function $f(x)$ that is imported into it) between the points $x = a$ and $x = b$. The input arguments are (Fun,a,b), where `Fun` is a dummy name that represents the imported function, and `a` and `b` are the endpoints of the domain. The function `funplot` also has a numerical output `xyout`, which is a 3×2 matrix with the values of x and $f(x)$ at the three points: $x = a$, $x = (a+b)/2$ and $x = b$. Note that in the program, the dummy function `Fun` has one input argument (x) and one output argument y, which are both vectors.

> A name for the function that is passed in.

```
function xyout=funplot(Fun,a,b)

% funplot makes a plot of the function Fun which is passed in
% when funplot is called in the domain [a, b].

% Input arguments are:
% Fun:   Function handle of the function to be plotted.

% a:   The first point of the domain.
% b:   The last point of the domain.

% Output argument is:
% xyout: The values of x and y at x=a, x=(a+b)/2, and x=b
% listed in A 3 by 2 matrix.

x=linspace(a,b,100);

y=Fun(x);
```

> Using the imported function to calculate $f(x)$ at 100 points.

```
xyout(1,1)=a; xyout(2,1)=(a+b)/2; xyout(3,1)=b;

xyout(1,2)=y(1);

xyout(2,2)=Fun((a+b)/2);

xyout(3,2)=y(100);
```

> Using the imported function to calculate $f(x)$ at the midpoint.

```
plot(x,y)
xlabel('x'), ylabel('y')
```

As an example, the function $f(x) = e^{-0.17x}x^3 - 2x^2 + 0.8x - 3$ over the domain [0.5, 4] is passed into the user-defined function `funplot`. This is done in two ways: first, by writing a user-defined function for $f(x)$, and then by writing $f(x)$ as an anonymous function.

Passing a user-defined function into a function function:

First, a user-defined function is written for $f(x)$. The function, named `Fdemo`, calculates $f(x)$ for a given value of x and is written using element-by-element operations.

```
function y=Fdemo(x)
y=exp(-0.17*x).*x.^3-2*x.^2+0.8*x-3;
```

Next, the function `Fdemo` is passed into the user-defined function function `fun-`

plot which is called in the Command Window. Note that a handle of the user-defined function Fdemo is entered (the handle is @Fdemo) for the input argument Fun in the user-defined function funplot.

```
>> ydemo=funplot(@Fdemo,0.5,4)
ydemo =
    0.5000   -2.9852
    2.2500   -3.5548
    4.0000    0.6235
```

> Enter a handle of the user-defined function Fdemo.

In addition to the display of the numerical output, when the command is executed, the plot shown in Figure 6-3 is also displayed in the Figure Window.

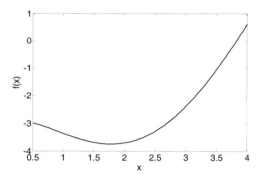

Figure 6-3: A plot of the function $f(x) = e^{-0.17x}x^3 - 2x^2 + 0.8x - 3$.

Passing an anonymous function into a function function:

To use an anonymous function, the function $f(x) = e^{-0.17x}x^3 - 2x^2 + 0.8x - 3$ first has to be written as an anonymous function, and then passed into the user-defined function funplot. The following shows how both of these steps are done in the Command Window. Note that the name of the anonymous function FdemoAnony is entered without the sign @ for the input argument Fun in the user-defined function funplot (since the name is already the handle of the anonymous function).

```
>> FdemoAnony=@(x) exp(-0.17*x).*x.^3-2*x.^2+0.8*x-3
FdemoAnony =
    @(x) exp(-0.17*x).*x.^3-2*x.^2+0.8*x-3
```

> Create an anonymous function for $f(x)$.

```
>> ydemo=funplot(FdemoAnony,0.5,4)
ydemo =
    0.5000   -2.9852
    2.2500   -3.5548
    4.0000    0.6235
```

> Enter a handle of the user-defined function FdemoAnony.

In, addition to the display of the numerical output in the Command Window, the plot shown in Figure 6-3 is displayed in the Figure Window.

6.9.2 Using a Function Name for Passing a Function into a Function Function

A second method for passing a function into a function function is by typing the name of the function that is being imported as a string in the input argument of the function function. The method can be used for importing user-defined functions. This method was used before the introduction of function handles. As mentioned, function handles are easier to use and more efficient and should be the preferred method to use. Importing user-defined functions by using their name is covered in the present edition of the book for the benefit of readers who need to understand programs that were written before MATLAB 7. New programs should use function handles.

When a user-defined function is imported by using its name, the value of the imported function inside the function function has to be calculated with the `feval` command. This is different when a function handle is used, which means that there is a difference in the way that the code in the function function is written that depends on how the imported function is passed in.

The `feval` command:

The `feval` (function evaluate) command evaluates the value of a function for a given value (or values) of the function's argument (or arguments). The format of the command is:

```
variable = feval('function name', argument value)
```

The value that is determined by `feval` can be assigned to a variable, or if the command is typed without an assignment MATLAB displays `ans =` and the value of the function.

- The function name is typed as string.

- The function can be a built-in or a user-defined function.

- If there is more than one input argument, the arguments are separated with commas.

- If there is more than one output argument, the variables on the left-hand side of the assignment operator are typed inside brackets and separated with commas.

Two examples of using the `feval` command with built-in functions are:

```
>> feval('sqrt',64)
ans =
     8
```

```
>> x=feval('sin',pi/6)
x =
    0.5000
```

The following shows the use of the `feval` command with the user-defined function `loan` that was created earlier in the chapter (Figure 6-2). This function has three input arguments and two output arguments.

```
                                          A $50,000 loan, 3.9% interest, 10 years.
>> [M,T]=feval('loan',50000,3.9,10)
M =
       502.22                             Monthly payment.
T =
     60266.47                             Total payment.
```

Writing a function function that accepts a function by typing its name as an input argument:

As already mentioned, when a user-defined function is imported by using its name, the value of the function inside the function function has to be calculated with the `feval` command. This is demonstrated in the following user-defined function function that is called `funplotS`. The function is the same as the function `funplot` from Section 6.9.1, except that the command `feval` is used for the calculations with the imported function.

A name for the function that is passed in.

```
function xyout=funplotS(Fun,a,b)

% funplotS makes a plot of the function Fun which is passed in
% when funplotS is called in the domain [a, b].

% Input arguments are:
% Fun: The function to be plotted. Its name is entered as
string expression.

% a:  The first point of the domain.
% b:  The last point of the domain.

% Output argument is:
% xyout: The values of x and y at x=a, x=(a+b)/2, and x=b
% listed in A 3 by 2 matrix.

x=linspace(a,b,100);

y=feval(Fun,x);    Using the imported function to calculate f(x) at 100 points.

xyout(1,1)=a; xyout(2,1)=(a+b)/2; xyout(3,1)=b;

xyout(1,2)=y(1);

xyout(2,2)=feval(Fun,(a+b)/2);    Using the imported function to
                                   calculate f(x) at the midpoint.
```

```
xyout(3,2)=y(100);
plot(x,y)
xlabel('x'), ylabel('y')
```

Passing a user-defined function into another function by using a string expression:

The following demonstrate how to pass a user-defined function into a function function by typing the name of the imported function as a string in the input argument. The function $f(x) = e^{-0.17x}x^3 - 2x^2 + 0.8x - 3$ from Section 6.9.1 that was created as a user-defined function named Fdemo is passed into the user-defined function funplotS. Note that the name Fdemo is typed in a string for the input argument Fun in the user-defined function funplotS.

```
>> ydemoS=funplotS('Fdemo',0.5,4)
ydemoS =
    0.5000    -2.9852
    2.2500    -3.5548
    4.0000     0.6235
```

The name of the imported function is typed as a string.

In, addition to the display of the numerical output in the Command Window, the plot shown in Figure 6-3 is displayed in the Figure Window.

6.10 SUBFUNCTIONS

A function file can contain more than one user-defined function. The functions are typed one after the other. Each function begins with a function definition line. The first function is called the primary function and the rest of the functions are called subfunctions. The subfunctions can be typed in any order. The name of the function file that is saved should correspond to the name of the primary function. Each of the functions in the file can call any of the other functions in the file. Outside functions, or programs (script files), can only call the primary function. Each of the functions in the file has its own workspace, which means that in each the variables are local. In other words, the primary function and the subfunctions cannot access each other variables (unless variables are declared to be global).

Subfunctions can help writing user-defined functions in an organized manner. The program in the primary function can be divided into smaller tasks, where each is carried out in a subfunction. This is demonstrated in Sample Problem 6-4.

Sample Problem 6-4: Average and standard deviation

Write a used-defined function that calculates the average and the standard deviation of a list of numbers. Use the function to calculate the average and the standard deviation of the following list of grades:

80 75 91 60 79 89 65 80 95 50 81

Solution

The average x_{ave} (mean) of a given set of n numbers $x_1, x_2, ..., x_n$ is given by:

$$x_{ave} = (x_1 + x_2 + ... + x_n)/n$$

The standard deviation is given by:

$$\sigma = \sqrt{\frac{\sum_{i=1}^{i=n} (x_i - x_{ave})}{n-1}}$$

A user-defined function, named `stat`, is written for solving the problem. To demonstrate the use of subfunctions, the function file includes `stat` as a primary function, and two subfunctions called `AVG` and `StandDiv`. The function `AVG` calculates x_{ave}, and the function `StandDiv` calculates σ. The subfunctions are called by the primary function. The following listing is saved as one function file called `stat`.

```
function [me SD] = stat(v)                    The primary function.
n=length(v);
me=AVG(v,n);
SD=StandDiv(v,me,n);


function av=AVG(x,num)                         Subfunction.
av=sum(x)/num;


function Sdiv=StandDiv(x,xAve,num)             Subfunction.
xdif=x-xAve;
xdif2=xdif.^2;
Sdiv= sqrt(sum(xdif2)/(num-1));
```

The user-defined function `stat` is then used in the Command Window for calculating the average and the standard deviation of the grades:

```
>> Grades=[80 75 91 60 79 89 65 80 95 50 81];
>> [AveGrade StanDeviation] = stat(Grades)
AveGrade =
    76.8182
StanDeviation =
    13.6661
```

6.11 NESTED FUNCTIONS

A nested function is a user-defined function that is written inside another user-defined function. The portion in the code that corresponds to the nested function starts with a function definition line and ends with an end statement. An end statement must be also entered at the end of the function that contains the nested function. (Normally, a user-defined function does not require a terminating end statement. However, an end statement is required if the function contains one or more nested functions.) Nested functions can also contain nested functions. Obviously, having many levels of nested functions can be confusing. This section considers only two levels of nested functions.

One nested function:

The format of a user-defined function A (called primary function) that contains one nested function B is:

```
function y=A(a1,a2)
. . . . . . .
      function z=B(b1,b2)
      . . . . . . .
      end
. . . . . . .
end
```

- Note the end statements at the end of functions B and A.

- The nested function B can access the workspace of the primary function A, and the primary function A can access the workspace of the function B. This means that a variable defined in the primary function A can be read and redefined in nested function B and vise versa.

- Function A can call function B, and function B can call function A.

Two (or more) nested functions at the same level:

The format of a user-defined function A (called primary function) that contains two nested functions B and C at the same level is:

```
function y=A(a1,a2)
. . . . . . .
      function z=B(b1,b2)
      . . . . . . .
      end
. . . . . . .
      function w=C(c1,c2)
      . . . . . . .
      end
. . . . . . .
end
```

- The three functions can access the workspace of each other.

- The three functions can call each other.

As an example, the following user-defined function (named statNest) that has two nested functions at the same level, solves the question in Sample Problem 6-4. Note that the nested functions are using variables (n and me) that are defined in the primary function.

```
function [me SD]=statNest(v)          The primary function.
n=length(v);
me=AVG(v);


    function av=AVG(x)                Nested function.
    av=sum(x)/n;
    end


    function Sdiv=StandDiv(x)         Nested function.
    xdif=x-me;
    xdif2=xdif.^2;
    Sdiv= sqrt(sum(xdif2)/(n-1));
    end


SD=StandDiv(v);
end
```

Using the user-defined function statNest in the Command Window for calculating the average of the given data gives:

```
>> Grades=[80 75 91 60 79 89 65 80 95 50 81];
>> [AveGrade StanDeviation] = statNest(Grades)
AveGrade =
   76.8182
StanDeviation =
   13.6661
```

Two levels of nested functions:

Two levels of nested functions are created when nested functions are written inside nested functions. The following shows an example for the format of a user-defined function with four nested functions in two levels.

```
function y=A(a1,a2)                    (Primary function A.)
.......
     function z=B(b1,b2)               (B is nested function in A.)
     .......
          function w=C(c1,c2)          (C is nested function in B.)
          .......
          end
     end
     function u=D(d1,d1)               (D is nested function in A.)
     .......
          function h=E(e1,e1)          (E is nested function in D.)
          .......
          end
     end
.......
end
```

The following rules apply to nested functions:

- A nested function can be called from a level above it. (In the preceding example function A can call B or D, but not C or E.)

- A nested function can be called from a nested function at the same level within the primary function. (In the preceding example function B can call D, and D can call B.)

- A nested function can be called from a nested function at any lower level.

- A variable defined in the primary function is recognized, and can be redefined, in a function that is nested at any level within the primary function.

- A variable defined in a nested function is recognized, and can be redefined, by any of the functions that contain the nested function.

6.12 EXAMPLES OF MATLAB APPLICATIONS

Sample Problem 6-5:   Exponential growth and decay

A model for exponential growth, or decay, of a quantity is given by:

$$A(t) = A_0 e^{kt}$$

where $A(t)$ and A_0 are the quantity at time t and time 0, respectively, and k is a constant unique to the specific application.

Write a user-defined function that uses this model to predict the quantity $A(t)$ at time t from knowing A_0 and $A(t_1)$ at some other time t_1. For function name and arguments use `At = expGD(A0,At1,t1,t)`, where the output argument `At` corresponds to $A(t)$, and the input arguments `A0,At1,t1,t` corresponds to A_0, $A(t_1)$, t_1, and t, respectively.

Use the function file in the Command Window for the following two cases:

a)   The population of Mexico was 67 million in the year 1980 and 79 million in 1986. Estimate the population in 2000.

b)   The half-life of a radioactive material is 5.8 years. How much of a 7-gram sample will be left after 30 years?

Solution

To use the exponential growth model, the value of the constant k has to be determined first by solving for k in terms of A_0, $A(t_1)$, and t_1:

$$k = \frac{1}{t_1} \ln \frac{A(t_1)}{A_0}$$

Once k is known, the model can be used to estimate the population at any time. The user-defined function that solves the problem is:

```
function At=expGD(A0,At1,t1,t)                    Function definition line.
% expGD calculates exponential growth and decay
% Input arguments are:
% A0: Quantity at time zero.
% At1: Quantity at time t1.
% t1: The time t1.
% t: time t.
% Output argument is:
% At: Quantity at time t.
k=log(At1/A0)/t1;                                 Determination of k.
At=A0*exp(k*t);                        Determination of A(t).
                                       (Assignment of value to output variable.)
```

Once the function is saved, it is used in the Command Window to solve the two cases. For case a) $A_0 = 67$, $A(t_1) = 79$, $t_1 = 6$, and $t = 20$:

```
>> expGD(67,79,6,20)
ans =
        116.03
```
Estimation of the population in the year 2000.

For case b) $A_0 = 7$, $A(t_1) = 3.5$ (since t_1 corresponds to half-life, which is the time required for the material to decay to half of its initial quantity), $t_1 = 5.8$, and $t = 30$.

```
>> expGD(7,3.5,5.8,30)
ans =
        0.19
```
The quantity of the material after 30 years.

Sample Problem 6-6:   Motion of a Projectile

Create a function file that calculates the trajectory of a projectile. The inputs to the function are the initial velocity and the angle at which the projectile is fired. The outputs from the function are the maximum height and distance. In addition, the function generates a plot of the trajectory. Use the function

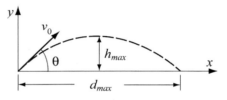

to calculate the trajectory of a projectile that is fired at a velocity of 230 m/s at an angle of 39°.

Solution

The motion of a projectile can be analyzed by considering the horizontal and vertical components. The initial velocity v_0 can be resolved into horizontal and vertical components:

$$v_{0x} = v_0\cos(\theta) \quad \text{and} \quad v_{0y} = v_0\sin(\theta)$$

In the vertical direction the velocity and position of the projectile are given by:

$$v_y = v_{0y} - gt \quad \text{and} \quad y = v_{0y}t - \frac{1}{2}gt^2$$

The time it takes the projectile to reach the highest point ($v_y = 0$) and the corresponding height are given by:

$$t_{hmax} = \frac{v_{0y}}{g} \quad \text{and} \quad h_{max} = \frac{v_{0y}^2}{2g}$$

The total flying time is twice the time it takes the projectile to reach the highest point, $t_{tot} = 2t_{hmax}$. In the horizontal direction the velocity is constant, and the position of the projectile is given by:

$$x = v_{0x}t$$

In MATLAB notation the function name and arguments are taken as: [hmax,dmax] = trajectory(v0,theta). The function file is:

```
function [hmax,dmax]=trajectory(v0,theta)    Function definition line.
% trajectory calculates the max height and distance of a
projectile, and makes a plot of the trajectory.
% Input arguments are:
% v0: initial velocity in (m/s).
% theta: angle in degrees.
% Output arguments are:
% hmax: maximum height in (m).
% dmax: maximum distance in (m).
% The function creates also a plot of the trajectory.
g=9.81;
v0x=v0*cos(theta*pi/180);
v0y=v0*sin(theta*pi/180);
thmax=v0y/g;
hmax=v0y^2/(2*g);
ttot=2*thmax;
dmax=v0x*ttot;
% Creating a trajectory plot
tplot=linspace(0,ttot,200);   Creating a time vector with 200 elements.
x=v0x*tplot;                   Calculating the x and y coordi-
y=v0y*tplot-0.5*g*tplot.^2;    nates of the projectile at each time.
plot(x,y)                      Note the element-by-element multiplication.
xlabel('DISTANCE (m)')
ylabel('HEIGHT (m)')
title('PROJECTILE''S TRAJECTORY')
```

After the function is saved, it is used in the Command Window for a projectile that is fired at a velocity of 230 m/s, and an angle of 39°.

```
>> [h d]=trajectory(230,39)
h =
   1.0678e+003
d =
   5.2746e+003
```

In addition, the following figure is created in the Figure Window:

6.13 Problems

1. The fuel efficiency of automobiles is measured in mi/Gal (miles per gallon) or in km/L (kilometers per liters). Write a MATLAB user-defined function that converts fuel efficiency values from mi/Gal (U.S. Gallons) to km/L. For the function name and arguments use `kmL=mgTOkm(mpg)`. The input argument `mpg` is the efficiency in mi/Gal, and the output argument `kmL` is the efficiency in km/L. Use the function in the Command Window to:

   a) Determine the fuel efficiency in km/L of a car that consumes 23 mi/Gal.
   b) Determine the fuel efficiency in km/L of a car that consumes 50 mi/Gal.

2. Write a user-defined MATLAB function, with two input and two output arguments, that determines the height in centimeters and mass in kilograms of a person from his height in inches and weight in pounds. For the function name and arguments use `[cm,kg] = STtoSI(in,lb)`. The input arguments are the height in inches and weight in pounds, and the output arguments are the height in centimeters and mass in kilograms. Use the function in the Command Window to:

   a) Determine in SI units the height and mass of a 5 ft. 11 in. person who weighs 181 lb.
   b) Determine your own height and weight in SI units.

3. Write a user-defined MATLAB function for the following math function:
$$y(x) = 0.9x^4 e^{-0.1x} - 15x^2 - 5x$$
The input to the function is x and the output is y. Write the function such that x can be a vector.

   a) Use the function to calculate $y(-2)$, and $y(4)$.
   b) Use the function to make a plot of the function $y(x)$ for $-3 \le x \le 5$.

4. Write a user-defined MATLAB function that converts speed given in units of kilometer per hour to speed in units of ft per second. For the function name and arguments use `ftps = kmphTOfps(kmh)`. The input argument is the speed in km/h., and the output argument is the speed in ft/s. Use the function to convert 70 km/h to units of ft/s.

5. Write a user-defined MATLAB function for the following function:

$$r(\theta) = \sin(3\theta)\cos\theta$$

The input to the function is θ (in radians) and the output is r. Write the function such that θ can be a vector.
a) Use the function to calculate $r(\pi/4)$, and $r(5\pi/2)$.
b) Use the function to plot (polar plot) $r(\theta)$ for $0 \le \theta \le 2\pi$.

6. Write a user-defined MATLAB function that calculates the local maximum or minimum of a quadratic function of the form: $f(x) = ax^2 + bx + c$. For the function name and arguments use `[x,y] = maxmin(a,b,c)`. The input arguments are the constants a, b, and c, and the output arguments are the coordinates x and y of the maximum or the minimum.

   Use the function to determine the maximum or minimum of the following functions:
a) $f(x) = 2x^2 + 9x - 20$
b) $f(x) = -3x^2 + 15x + 50$

7. The monthly payment M of a loan of amount P for N years and an annual interest rate r (in %) can be calculated by the formula:

$$M = P\frac{\dfrac{r}{1200}}{1 - \left(1 + \dfrac{r}{1200}\right)^{-N}}$$

Write a MATLAB user-defined function that calculates the monthly payment of a loan. For the function name and arguments use `M = amort(P,r,N)`. The inputs arguments are P the loan amount, r the annual interest rate in percent, and N the length of the loan in years. The output M is the amount of the monthly payment. Use the function to calculate the monthly payment of a 15-year mortgage of $260,000 with annual interest rate of 6.75%.

8. The weight W of a ring in a shape of a torus with an inner radius r and a diameter d is given by:

$$W = \gamma\frac{1}{4}\pi^2(2r + d)d^2$$

where γ is the specific weight of the ring material. Write an anonymous function that calculates the weight of the ring. The function should have three input arguments r, d, and γ. Use the anonymous function to calculate the weight of a gold ring ($\gamma = 0.696$ lb/in^3) with $r = 0.6$ in., $d = 0.092$ in.

9. Write a user-defined MATLAB function that determines the area of a triangle when the lengths of the sides are given. For the function name and arguments use [Area] = triangle(a,b,c). Use the function to determine the area of triangles with the following sides:
   a) $a = 10$, $b = 15$, $c = 7$.
   b) $a = 6$, $b = 8$, $c = 10$.
   c) $a = 200$, $b = 75$, $c = 250$.

10. Write a user-defined MATLAB function that determines the unit vector in the direction of the line that connects two points (A and B) in space. For the function name and arguments use n = unitvec(A,B). The input to the function are two vectors A and B, each of which has three elements which are the Cartesian coordinates of the corresponding point. The output is a vector with the three components of the unit vector in the direction from A to B. Use the function to determine the following unit vectors:
   a) In the direction from point (1.5, 2.1, 4) to point (11, 15, 9).
   b) In the direction from point (–11, 3, -2) to point (-13, -4, -5).
   c) In the direction from point (1, 0, 1) to point (0, 1, 1).

11. In polar coordinates a two-dimensional vector is given by its radius and angle (r, θ). Write a user-defined MATLAB function that adds two vectors that are given in polar coordinates. For the function name and arguments use:
    [r th] = AddVecPol(r1,th1,r2,th2)
    where the input arguments are (r_1, θ_1) and
    (r_2, θ_2), and the output arguments are the radius and angle of the result. Use the function to carry out the following additions:
    a) $r_1 = (5, 23°)$, $r_2 = (12, 40°)$.   b) $r_1 = (6, 80°)$, $r_2 = (15, 125°)$.

12. Write a user-defined MATLAB function that gives a random integer number within a range between two numbers. For the function name and arguments use n = randint(a,b), where the two input arguments a and b are the two numbers and the output argument n is the random number.
    Use the function in the Command Window for the following:
    a) Generate a random number between 1 and 49.
    b) Generate a random number between –35 and –2.

13. Write a user-defined MATLAB function that calculates the determinant of a 3×3 matrix by using the formula:

$$det = A_{11} \begin{vmatrix} A_{22} & A_{23} \\ A_{32} & A_{33} \end{vmatrix} - A_{12} \begin{vmatrix} A_{21} & A_{23} \\ A_{31} & A_{33} \end{vmatrix} + A_{13} \begin{vmatrix} A_{21} & A_{22} \\ A_{31} & A_{32} \end{vmatrix}$$

For the function name and arguments use d3 = det3by3(A), where the input argument A is the matrix and the output argument d3 is the value of the determinant. Write the code of det3by3 such that it has a subfunction that calculates the 2×2 determinant. Use det3by3 for calculating the determinants of:

a) $\begin{bmatrix} 1 & 3 & 2 \\ 6 & 5 & 4 \\ 7 & 8 & 9 \end{bmatrix}$ b) $\begin{bmatrix} -2.5 & 7 & 1 \\ 5 & -3 & -2.6 \\ 4 & 2 & -1 \end{bmatrix}$

14. Write a user-defined MATLAB function that calculates a student's final grade in a course using the scores from three midterm exams, a final exam, and six homework assignments. The midterms are graded on a scale from 0 to 100, and are each 15% of the final grade. The final exam is graded on a scale from 0 to 100, and is 45% of the final grade. The six homework assignments are graded each on a scale from 0 to 10. The homework assignments together are 10% of the final grade.

   For the function name and arguments use g = fgrade(R). The input argument R is a matrix in which the elements in each row are the grades of one student. The first six columns are the homework grades (numbers between 0 and 10), the next three columns are the midterm grades (numbers between 0 and 100), and the last column is the final exam grade (a number between 0 and 100). The output from the function, g, is a column vector with the final grades for the course. Each row has the final grade of the student with the grades in the corresponding row of the matrix R.

   The function can be used to calculate the grades of any number of students. For one student the matrix R has one row. Use the function in the following cases:

   a) Use the Command Window to calculate the grade of one student with the following grades: (8, 9, 6, 10, 9, 8, 76, 86, 91, 80).

   b) Write a program in a script file. The program asks the user to enter the students' grades in an array (each student a row). The program then calculates the final grades by using the function fgrade. Run the script file in the Command Window to calculate the grades of the following four students:
   Student A: 8, 10, 6, 9, 10, 9, 91, 71, 81, 85.
   Student B: 5, 5, 6, 1, 8, 6, 59, 72, 66, 59.
   Student C: 6, 8, 10, 4, 5, 9, 55, 65, 75 78.
   Student D: 7, 7, 8, 8, 9, 8, 83, 82, 81 84.

15. When n electrical resistors are connected in parallel, their equivalent resistance R_{Eq} can be determined from:

$$\frac{1}{R_{Eq}} = \frac{1}{R_1} + \frac{1}{R_2} + \dots + \frac{1}{R_n}$$

Write a user-defined MATLAB function that calculates R_{Eq}. For the function name and arguments use REQ = req(R). The input to the function is a vector in which each element is a resistor value, and the output from the function is R_{Eq}. Use the function to calculate the equivalent resistance when the following resistors are connected in parallel: 50Ω, 75Ω, 300Ω, 60Ω, 500Ω, 180Ω, and 200Ω.

16. Write a user-defined function that determines the coordinate y_C of the centroid of the U-shaped cross-sectional area shown in the figure. For the function name and arguments use yc = centroidU(w,h,t), where the input arguments w, h, and t are the dimensions shown in the figure, and the output argument yc is the coordinate y_C.

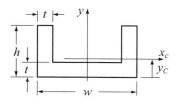

Use the function to determine y_C for an area with $w = 250$ mm, $h = 160$ mm, and $t = 26$ mm.

17. A two-dimensional state of stress at a point in a loaded material is defined by three components of stresses σ_{xx}, σ_{yy}, and τ_{xy}. The maximum and minimum normal stresses (principal stresses) at the point, σ_{max} and σ_{min}, are calculated from the stress components by:

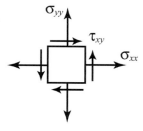

$$\sigma_{\substack{max \\ min}} = \frac{\sigma_{xx} + \sigma_{yy}}{2} \pm \sqrt{\left(\frac{\sigma_{xx} - \sigma_{yy}}{2}\right)^2 + \tau_{xy}^2}$$

Write a user-defined MATLAB function that determines the principal stresses from the stress components. For the function name and arguments use [Smax,Smin] = princstress(Sxx,Syy,Sxy). The input arguments are the three stress components, and the output arguments are the maximum and minimum stresses.

Use the function to determine the principal stresses for the following states of stress:

a) $\sigma_{xx} = -190$ MPa, $\sigma_{yy} = 145$ MPa, and $\tau_{xy} = 110$ MPa.

b) $\sigma_{xx} = 14$ ksi, $\sigma_{yy} = -15$ ksi, and $\tau_{xy} = 8$ ksi.

18. The area moment of inertia I_{x_o} of a rectangle about the

    axis x_o that passes through its centroid is: $I_{x_o} = \frac{1}{12}bh^2$.

    The moment of inertia about an axis x that is parallel to

    x_o is given by: $I_x = I_{x_o} + Ad_x^2$, where A is the area of the

    rectangle, and d_x is the distance between the two axes.

    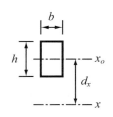

    Write a MATLAB user-defined function that determines the area moment of inertia I_{x_C} of an "U" beam about the axis that passes through its centroid (see drawing). For the function name and arguments use Ixc = IxcBeam(w,h,t). The input arguments to the function are the width w, the height h, and

    the thickness t. The output argument Ixc is I_{x_C}. For finding the coordinate y_C of the of the centroid use the user-defined function centroidU from Problem 16 as a subfunction inside IxcBeam.

    (The moment of inertia of a composite area is obtained by dividing the area into parts and adding the moment of inertia of the parts.)

    Use the function to determine the moment of inertia of an "I" beam with $w = 320$ mm, $h = 180$ mm, and $t = 32$ mm.

19. In a low-pass RC filter (a filter that passes signals with low frequencies), the ratio of the magnitude of the voltages is given by:

    $$RV = \left|\frac{V_o}{V_i}\right| = \frac{1}{\sqrt{1+(\omega RC)^2}}$$

    where ω is the frequency of the input signal.

    Write a user-defined MATLAB function that calculates the magnitude ratio. For the function name and arguments use RV = lowpass(R,C,w). The input arguments are: R the size of the resistor in Ω (ohms), C the size of the capacitor in F (Farads), and w the frequency of the input signal in rad/s. Write the function such that w can be a vector.

    Write a program in a script file that uses the lowpass function to generate a plot of RV as a function of ω for $10^{-2} \leq \omega \leq 10^6$ rad/s. The plot has a logarithmic scale on the horizontal axis (ω). When the script file is executed, it asks the user to enter the values of R and C. Label the axes of the plot.

    Run the script file with $R = 1200\,\Omega$, and $C = 8\,\mu$F.

20. A band-pass filter passes signals with fre-
    quencies within a certain range. In this filter
    the ratio of the magnitude of the voltages is
    given by:

$$RV = \left|\frac{V_o}{V_i}\right| = \frac{\omega RC}{\sqrt{(1 - \omega^2 LC)^2 + (\omega RC)^2}}$$

where ω is the frequency of the input signal.
Write a user-defined MATLAB function that calculates the magnitude ratio.
For the function name and arguments use RV = bandpass (R, C, L, w). The
input arguments are: R the size of the resistor in Ω (ohms), C the size of the
capacitor in F (Farads), L the inductance of the coil in H (Henrys), and w the
frequency of the input signal in rad/s. Write the function such that w can be a
vector.

    Write a program in a script file that uses the bandpass function to gen-
erate a plot of RV as a function of ω for $10^{-2} \le \omega \le 10^7$ rad/s. The plot has a
logarithmic scale on the horizontal axis (ω). When the script file is executed,
it asks the user to enter the values of R, L, and C. Label the axes of the plot.

    Run the script file for the following two cases:

a) $R = 1100\,\Omega$, and $C = 9\,\mu F$, $L = 7\,mH$.
b) $R = 500\,\Omega$, and $C = 300\mu F$, $L = 400mH$.

21. The dew point temperature T_d can be calculated (approximately) from the rel-
    ative humidity RH and the actual temperature T by (http://www.paroscien-
    tific.com/dewpoint.htm):

$$T_d = \frac{b \, f(T, RH)}{a - f(T, RH)} \quad \text{with} \quad f(T, RH) = \frac{aT}{b + T} + \ln\left(\frac{RH}{100}\right)$$

where the temperature is in degrees Celsius, RH in %, $a = 17.27$, and
$b = 237.7\,^{\circ}C$.

    Write a user-defined MATLAB function that calculates the dew point
temperature for a given temperature and relative humidity. For the function
name and arguments use Td = dewpoint (T, bRH), where the two input
arguments T and RH are the temperature and relative humidity, respectively,
and the output argument Td is the dew point temperature. Use the user-
defined function dewpoint for calculating the dew point temperature for the
following cases:

a) $T = 15\,^{\circ}C$, $RH = 40\%$.
b) $T = 35\,^{\circ}C$, $RH = 80\%$.

Chapter 7
Programming in MATLAB

A computer program is a sequence of computer commands. In a simple program the commands are executed one after the other in the order that they are typed. In this book, for example, all the programs that have been presented so far, in script or function files, are simple programs. Many situations, however, require more sophisticated programs in which commands are not necessarily executed in the order that they are typed, or that different commands (or groups of commands) are executed when the program runs with different input variables. For example, a computer program that calculates the cost of mailing a package for the post office uses different mathematical expressions to calculate the cost depending on the weight and size of the package, the content (books are less expensive to mail), and the type of service (airmail, ground, etc.). In other situations there might be a need to repeat a sequence of commands several times within a program. For example, programs that solve equations numerically repeat a sequence of calculations until the error in the answer is smaller than some measure.

MATLAB provides several tools that can be used to control the flow of a program. Conditional statements (Section 7.2) and the `switch` structure (Section 7.3), make it possible to skip commands or to execute specific groups of commands in different situations. `For` loops and `while` loops (Section 7.4) make it possible to repeat a sequence of commands several times.

It is obvious that changing the flow of a program requires some kind of decision-making process within the program. The computer must decide whether to execute the next command or to skip one or more commands and continue at a different line in the program. The program makes these decisions by comparing values of variables. This is done by using relational and logical operators which are explained in Section 7.1.

It should also be noted that function files (Chapter 6) can be used in programming. A function file is a subprogram. When the program reaches the command line that has the function, it provides input to the function, and "waits" for the results. The function carries out the calculations, transfers the results back to the program that "called" the function, which then continues to the next command.

7.1 RELATIONAL AND LOGICAL OPERATORS

A relational operator compares two numbers by determining whether a comparison statement (e.g. 5 < 8) is true or false. If the statement is true, it is assigned a value of 1. If the statement is false, it is assigned a value of 0. A logical operator examines true/false statements and produces a result which is true (1) or false (0) according to the specific operator. For example, the logical AND operator gives 1 only if both statements are true. Relational and logical operators can be used in mathematical expressions, and, as will be shown in this chapter, are frequently used in combination with other commands, to make decisions that control the flow of a computer program.

Relational Operators:

Relational operators in MATLAB are:

Relational Operator	Description
<	Less than.
>	Greater than.
<=	Less than or equal to.
>=	Greater than or equal to.
==	Equal to.
~=	Not Equal to.

Note that the "equal to" relational operator consists of two = signs (with no space between them), since one = sign is the assignment operator. Also, in other relational operators that consist of two characters there is no space between the characters (<=, >=, ~=)

- Relational operators are used as arithmetic operators within a mathematical expression. The result can be used in other mathematical operations, in addressing arrays, and together with other MATLAB commands (e.g. if) to control the flow of a program.

- When two numbers are compared, the result is 1 (logical true) if the comparison, according to the relational operator, is true, and 0 (logical false) if the comparison is false.

- If two scalars are compared, the result is a scalar 1 or 0. If two arrays are compared (only arrays with the same size can be compared), the comparison is done *element-by-element*, and the result is a logical array of the same size with 1's and 0's according to the outcome of the comparison at each address.

- If a scalar is compared with an array, the scalar is compared with every element of the array, and the result is a logical array with 1's and 0's according to the outcome of the comparison of each element.

Some examples are:

```
>> 5>8
```
Checks if 5 is larger than 8.
```
ans =
     0
```
Since the comparison is false (5 is not larger than 8) the answer is 0.
```
>> a=5<10
```
Checks if 5 is smaller than 10, and assigns the answer to a.
```
a =
     1
```
Since the comparison is true (5 is smaller than 10) the number 1 is assigned to a.

```
>> y=(6<10)+(7>8)+(5*3= =60/4)
```
Using relational operators in math expression.

Equal to 1 since 6 is smaller than 10.

Equal to 0 since 7 is not larger than 8.

Equal to 1 since 5*3 is equal to 60/4.

```
y =
     2
>> b=[15 6 9 4 11 7 14]; c=[8 20 9 2 19 7 10];
```
Define vectors b and c.

```
>> d=c>=b
```
Checks which c elements are larger or equal to b elements.
```
d =
     0     1     1     0
```
Assigns 1 where an element of c is larger than or equal to an element of b.
```
>> b == c
```
Checks which b elements are equal to c elements.
```
ans =
     0     0     1     0     0     1     0
>> b~=c
```
Checks which b elements are not equal to c elements.
```
ans =
     1     1     0     1     1     0     1
>> f=b-c>0
```
Subtracts c from b and then checks which elements are larger than zero.
```
f =
     1     0     0     1     0     0     1
>> A=[2 9 4; -3 5 2; 6 7 -1]
```
Define a 3 × 3 matrix A.
```
A =
     2     9     4
    -3     5     2
     6     7    -1
```
Checks which elements in A are smaller or equal to 2. Assigns the results to matrix B.
```
>> B=A<=2
```

```
B =
     1     0     0
     1     0     1
     0     0     1
```

- The results of a relational operation with vectors, which are vectors with 0's and 1's, are called logical vectors and can be used for addressing vectors. When a logical vector is used for addressing another vector, it extracts from that vector the elements in the positions where the logical vector has 1's. For example:

```
>> r = [8 12 9 4 23 19 10]                         Define a vector r.
r =
     8    12     9     4    23    19    10
>> s=r<=10             Checks which r elements are smaller than or equal to 10.
s =                         A logical vector s with 1's at positions where
     1     0     1     1     elements of r are smaller than or equal to 10.
>> t=r(s)               Use s for addresses in vector r to create vector t.
t =                              Vector t consists of elements of
     8     9     4    10          r in positions where s has 1's.
>> w=r(r<=10)                    The same can be done in one step.
w =
     8     9     4    10
```

- Numerical vectors and arrays with the numbers 0's and 1's are not the same as logical vectors and arrays with 0's and 1's. Numerical vectors and arrays can not be used for addressing. Logical vectors and arrays, however, can be used in arithmetic operations. The first time a logical vector or an array is used in arithmetic operations it is changed to a numerical vector or array.

- Order of precedence: In a mathematical expression that includes relational and arithmetic operations, the arithmetic operations (+, –, *, /, \) have precedence over relational operations. The relational operators themselves have equal precedence and are evaluated from left to right. Parentheses can be used to alter the order of precedence. Examples are:

```
>> 3+4<16/2                                   + and / are executed first.
ans =                                  The answer is 1 since 7 < 8 is true.
     1
>> 3+(4<16)/2          4 < 16 is executed first, and is equal to 1, since it is true.
ans =                                      3.5 is obtained from 3 + 1/2.
     3.5000
```

Logical Operators:

Logical operators in MATLAB are:

Logical Operator	Name	Description
& Example: A&B	AND	Operates on two operands (A and B). If both are true, the result is true (1), otherwise the result is false (0).
| Example: A|B	OR	Operates on two operands (A and B). If either one, or both are true, the result is true (1), otherwise (both are false) the result is false (0).
~ Example: ~A	NOT	Operates on one operand (A). Gives the opposite of the operand. True (1) if the operand is false, and false (0) if the operand is true.

- Logical operators have numbers as operands. A nonzero number is true, and a zero number is false.

- Logical operators (like relational operators) are used as arithmetic operators within a mathematical expression. The result can be used in other mathematical operations, in addressing arrays, and together with other MATLAB commands (e.g. if) to control the flow of a program.

- Logical operators (like relational operators) can be used with scalars and arrays.

- The logical operations AND and OR can have both operands as scalars, arrays, or one array and one a scalar. If both are scalars, the result is a scalar 0 or 1. If both are arrays they must be of the same size and the logical operation is done *element-by-element*. The result is an array of the same size with 1's and 0's according to the outcome of the operation at each position. If one operand is a scalar and the other is an array, the logical operation is done between the scalar and each of the elements in the array and the outcome is an array of the same size with 1's and 0's.

- The logical operation NOT has one operand. When it is used with a scalar the outcome is a scalar 0 or 1. When it is used with an array, the outcome is an array of the same size with 1's in positions where the array has nonzero numbers and 0's in positions where the array has zeros.

The following are some examples:

```
>> 3&7
```

3 AND 7.

```
ans =
    1
```
3 and 7 are both true (nonzero), so the outcome is 1.

```
>> a=5|0
```
5 OR 0 (assign to variable a).

```
a =
    1
```
1 is assigned to a since at least one number is true (nonzero).

```
>> ~25
```
NOT 25.

```
ans =
    0
```
The outcome is 0 since 25 is true (nonzero) and the opposite is false.

```
>> t=25*((12&0)+(~0)+(0|5))
```
Using logical operators in math expression.

```
t =
    50
```

```
>> x=[9 3 0 11 0 15]; y=[2 0 13 -11 0 4];
```
Define two vectors x and y.

```
>> x&y
ans =
    1    0    0    1
```
The outcome is a vector with 1 in every position that both x and y are true (nonzero elements), and 0's otherwise.

```
>> z=x|y
z =
    1    1    1    1
```
The outcome is a vector with 1 in every position that either or both x and y are true (nonzero elements), and 0's otherwise.

```
>> ~(x+y)
ans =
    0    0    0    1
```
The outcome is a vector with 0 in every position that the vector x + y is true (nonzero elements), and 1 in every position that x + y is false (zero element).

Order of precedence:

Arithmetic, relational, and logical operators can all be combined in mathematical expressions. When an expression has such a combination, the result depends on the order in which the operations are carried out. The following is the order used by MATLAB:

Precedence	Operation
1 (highest)	Parentheses (If nested parentheses exist, inner have precedence).
2	Exponentiation.
3	Logical NOT (~).
4	Multiplication, division.
5	Addition, subtraction.
6	Relational operators (>, <, >=, <=, ==, ~=).
7	Logical AND (&).
8 (lowest)	Logical OR (	).

If two or more operations have the same precedence, the expression is executed in order from left to right.

It should be pointed out here that the order shown above is used since MAT-LAB 6. Previous versions of MATLAB have a slightly different order (& did not have precedence over |), and the user must be careful. Compatibility problems between different versions of MATLAB can be avoided by using parentheses even when they are not required.

The following are examples of expressions that include arithmetic, relational, and logical operators:

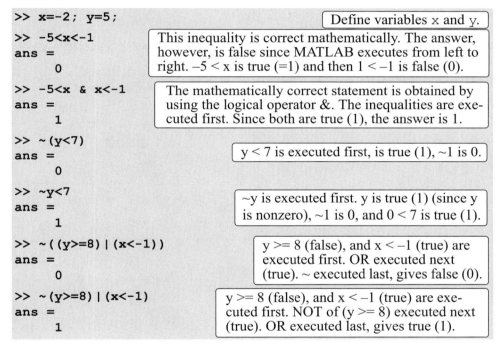

```
>> x=-2; y=5;
```
Define variables x and y.

```
>> -5<x<-1
ans =
     0
```
This inequality is correct mathematically. The answer, however, is false since MATLAB executes from left to right. −5 < x is true (=1) and then 1 < −1 is false (0).

```
>> -5<x & x<-1
ans =
     1
```
The mathematically correct statement is obtained by using the logical operator &. The inequalities are executed first. Since both are true (1), the answer is 1.

```
>> ~(y<7)
ans =
     0
```
y < 7 is executed first, is true (1), ~1 is 0.

```
>> ~y<7
ans =
     1
```
~y is executed first. y is true (1) (since y is nonzero), ~1 is 0, and 0 < 7 is true (1).

```
>> ~((y>=8)|(x<-1))
ans =
     0
```
y >= 8 (false), and x < −1 (true) are executed first. OR executed next (true). ~ executed last, gives false (0).

```
>> ~(y>=8)|(x<-1)
ans =
     1
```
y >= 8 (false), and x < −1 (true) are executed first. NOT of (y >= 8) executed next (true). OR executed last, gives true (1).

Built-in Logical Functions:

MATLAB has built-in functions that are equivalent to the logical operators. These functions are:

and(A,B)   equivalent to A&B
or(A,B)   equivalent to A|B
not(A)   equivalent to ~A

In addition, MATLAB has other logical built-in functions, some of which are described in the following table:

Function	Description	Example

xor(a,b)	Exclusive or. Returns true (1) if one operand is true and the other is false.	`>> xor(7,0)`
		`ans =`
		    `1`
		`>> xor(7,-5)`
		`ans =`
		    `0`

all(A)	Returns 1 (true) if all elements in a vector A are true (nonzero). Returns 0 (false) if one or more elements are false (zero).
	If A is a matrix, treats columns of A as vectors, returns a vector with 1's and 0's.
		`>> A=[6 2 15 9 7 11];`
		`>> all(A)`
		`ans =`
		    `1`
		`>> B=[6 2 15 9 0 11];`
		`>> all(B)`
		`ans =`
		    `0`

any(A)	Returns 1 (true) if any element in a vector A is true (nonzero). Returns 0 (false) if all elements are false (zero).
	If A is a matrix, treats columns of A as vectors, returns a vector with 1's and 0's.
		`>> A=[6 0 15 0 0 11];`
		`>> any(A)`
		`ans =`
		    `1`
		`>> B = [0 0 0 0 0 0];`
		`>> any(B)`
		`ans =`
		    `0`

find(A)
 	
find(A>d)
	If A is a vector, returns the indices of the nonzero elements.
	
	If A is a vector, returns the address of the elements that are larger than d (any relational operator can be used).
		`>> A=[0 9 4 3 7 0 0 1 8];`
		`>> find(A)`
		`ans =`
		    `2    3    4`
		`5    8    9`
		`>> find(A>4)`
		`ans =`
		    `2    5    9`

The operation of the four logical operators, and, or, xor, and not can be summarized in a truth table:

INPUT		OUTPUT				
A	B	AND A&B	OR A|B	XOR (A,B)	NOT ~A	NOT ~B
false	false	false	false	false	true	true
false	true	false	true	true	true	false
true	false	false	true	true	false	true
true	true	true	true	false	false	false

Sample Problem 7-1:   Analysis of temperature data

The following were the daily maximum temperatures (in °F) in Washington DC during the month of April, 2002: 58 73 73 53 50 48 56 73 73 66 69 63 74 82 84 91 93 89 91 80 59 69 56 64 63 66 64 74 63 69, (data from the U.S. National Oceanic and Atmospheric Administration). Use relational and logical operations to determine the following:

a)   The number of days the temperature was above 75°.

b)   The number of days the temperature was between 65° and 80°.

c)   The days of the month that the temperature was between 50° and 60°.

Solution

In the script file below the temperatures are entered in a vector. Relational and logical expressions are then used to analyze the data.

```
T=[58 73 73 53 50 48 56 73 73 66 69 63 74 82 84 ...
   91 93 89 91 80 59 69 56 64 63 66 64 74 63 69];
Tabove75=T>=75;
```
A vector with 1's at addresses where T >= 75.

```
NdaysTabove75=sum(Tabove75)
```
Add all the 1's in the vector Tabove75.

```
Tbetween65and80=(T>=65)&(T<=80);
```
A vector with 1's at addresses where T >= 65 and T <= 80.

```
NdaysTbetween65and80=sum(Tbetween65and80)
```
Add all the 1's in the vector Tbetween65and80.

```
datesTbetween50and60=find((T>=50)&(T<=60))
```
The function find returns the address of the elements in T that have values between 50 and 60.

The script file (saved as Exp7_1) is executed in the Command Window:

```
>> Exp7_1
NdaysTabove75 =
     7
NdaysTbetween65and80 =
    12
datesTbetween50and60 =
     1      4      5      7     21     23
```

For 7 days the temp was above 75.

For 12 days the temp was between 65 and 80.

Dates of the month with temp between 50 and 60.

7.2 CONDITIONAL STATEMENTS

A conditional statement is a command that allows MATLAB to make a decision of whether to execute a group of commands that follow the conditional statement, or to skip these commands. In a conditional statement a conditional expression is stated. If the expression is true, a group of commands that follow the statement is executed. If the expression is false, the computer skips the group. The basic form of a conditional statement is:

if conditional expression consisting of relational and/or logical operators.

Examples:

```
if a < b
if c >= 5
if a == b
if a ~= 0
if (d<h) & (x>7)
if (x~=13) | (y<0)
```

All the variables must have assigned values.

• Conditional statements can be a part of a program written in a script file or a function file.

• As shown below, for every if statement there is an end statement.

The if statement is commonly used in three structures, if-end, if-else-end, and if-elseif-else-end, which are described next.

7.2.1 The if-end Structure

The if-end structure of the conditional statement is shown schematically in Figure 7-1. The figure shows how the commands are typed in the program, and a flowchart that symbolically shows the flow, or the sequence, in which the commands are executed. As the program executes, it reaches the if statement. If the

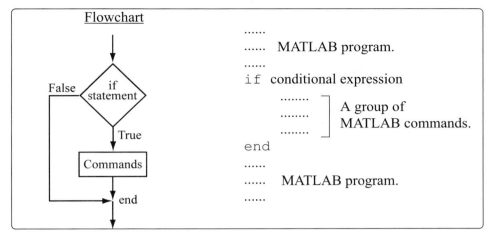

Figure 7-1: The structure of the `if-end` **conditional statement.**

conditional expression in the `if` statement is true (1), the program continues to execute the commands that follow the `if` statement all the way down to the `end` statement. If the conditional expression is false (0), the program skips the group of commands between the `if` and the `end`, and continues with the commands that follow the `end`.

The words `if` and `end` appear on the screen in blue, and the commands between the `if` statement and the `end` statement are automatically indented (they don't have to be indented), which makes the program easier to read. An example where the `if-end` statement is used in a script file is shown in Example 7-2.

Sample Problem 7-2:   Calculating worker's pay

A worker is paid according to his hourly wage up to 40 hours, and 50% more for overtime. Write a program in a script file that calculates the pay to a worker. The program asks the user to enter the number of hours and the hourly wage. The program then displays the pay.

Solution

The program in a script file is shown below. The program first calculates the pay by multiplying the number of hours by the hourly wage. Then an `if` statement checks whether the number of hours is greater than 40. If yes, the next line is executed and the extra pay for the hours above 40 is added. If not, the program skips to the `end`.

```
t=input('Please enter the number of hours worked  ');
h=input('Please enter the hourly wage in $  ');
Pay=t*h;
if t>40
```

```
    Pay=Pay+(t-40)*0.5*h;
end
fprintf('The worker''s pay is  $ %5.2f',Pay)
```

Application of the program (in the Command Window) for two cases is shown below (the file was saved as Workerpay):

```
>> Workerpay
Please enter the number of hours worked  35
Please enter the hourly wage in $  8
The worker's pay is  $ 280.00
>> Workerpay
Please enter the number of hours worked  50
Please enter the hourly wage in $  10
The worker's pay is  $ 550.00
```

7.2.2 The if-else-end Structure

This structure provides a means for choosing one group of commands, out of a possible two groups, for execution. The if-else-end structure is shown in Figure 7-2.The figure shows how the commands are typed in the program, and a flowchart that illustrates the flow, or the sequence, in which the commands are

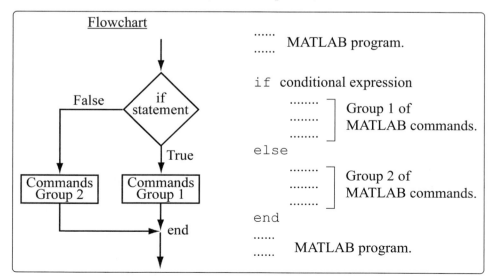

Figure 7-2: The structure of the if-else-end **conditional statement.**

executed. The first line is an `if` statement with a conditional expression. If the conditional expression is true, the program executes group 1 of commands between the `if` and the `else` statements and then skips to the end. If the conditional expression is false, the program skips to the `else`, and then executes group 2 of commands between the `else` and the `end`.

The following example uses the `if-else-end` structure in a function file.

Sample Problem 7-3: Water level in water tower

The tank in a water tower has the geometry shown in the figure (the lower part is a cylinder and the upper part is an inverted frustum cone). Inside the tank there is a float that indicates the level of the water. Write a user-defined function file that determines the volume of the water in the tank from the position (height h) of the float. The input to the function is the value of h in m, and the output is the volume of the water in m³.

Solution

For $0 \le h \le 19$ m the volume of the water is given by the volume of a cylinder with height h: $V = \pi 12.5^2 h$.

For $19 < h \le 33$ m the volume of the water is given by adding the volume of a cylinder with $h = 19$ m, and the volume of the water in the cone:

$$V = \pi 12.5^2 \cdot 19 + \frac{1}{3}\pi(h-19)(12.5^2 + 12.5 \cdot r_h + r_h^2)$$

where r_h is: $r_h = 12.5 + \frac{10.5}{14}(h-19)$

The function, shown below, is named $v = \texttt{watervol(h)}$.

```
function v = watervol(h)
% watervol calculates the volume of the water in the water
tower.
% The input is the water level in meters.
% The output is the water volume in cubic meters.

if h<=19
    v = pi*12.5^2*h;
else
    rh=12.5+10.5*(h-19)/14;
    v=pi*12.5^2*19+pi*(h-19)*(12.5^2+12.5*rh+rh^2)/3;
```

```
end
```

Two examples of using the function (in the Command Window) are given next:

```
>> watervol(8)
ans =
   3.9270e+003
>> VOL=watervol(25.7)
VOL =
   1.4115e+004
```

7.2.3 The if-elseif-else-end Structure

The `if-elseif-else-end` structure is shown in Figure 7-3. The figure shows how the commands are typed in the program, and a flowchart that illustrates the flow, or the sequence, in which the commands are executed. This structure includes two conditional statements (`if` and `elseif`) which make it possible to select one out of three groups of commands for execution. The first line is an `if` statement with a conditional expression. If the conditional expression is true, the program executes group 1 of commands between the `if` and the

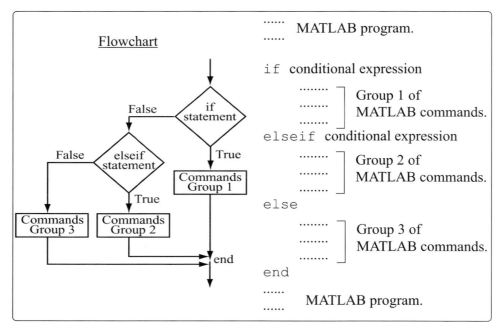

Figure 7-3: The structure of the `if-elseif-else-end` **conditional statement.**

`elseif` statement and then skips to the `end`. If the conditional expression in the `if` statement is false, the program skips to the `elseif` statement. If the conditional expression in the `elseif` statement is true the program executes group 2 of commands between the `elseif` and the `else` and then skips to the `end`. If the conditional expression in the `elseif` statement is false the program skips to the `else` and executes group 3 of commands between the `else` and the `end`.

It should be pointed out here that several `elseif` statements and associated groups of commands can be added. In this way more conditions can be included. Also, the `else` statement is optional. This means that in the case of several `elseif` statements and no `else` statement, if any of the conditional statements is true the associated commands are executed, but otherwise nothing is executed.

7.3 *THE* `switch-case` *STATEMENT*

The `switch-case` statement is another method that can be used to affect the flow of a program. It provides a means for choosing one group of commands for execution out of several possible groups. The structure of the statement is shown in Figure 7-4.

```
......        MATLAB program.
......

switch  switch expression
     case  value1
        ........      ] Group 1 of commands.
        ........
     case  value2
        ........      ] Group 2 of commands.
        ........
     case  value3
        ........      ] Group 3 of commands.
        ........
     otherwise
        ........      ] Group 4 of commands.
        ........
end
......        MATLAB program.
......
```

Figure 7-4: The structure of a `switch-case` **statement.**

- The first line is the `switch` command which has the form:

| switch   switch expression |

The switch expression can be a scalar or a string. Usually it is a variable that has an assigned scalar or a string. It can also be, however, a mathematical expression that includes preassigned variables and can be evaluated.

- Following the `switch` command there are one or several `case` commands. Each has a value (can be a scalar or a string) next to it (value1, value2, etc.) and associated group of commands below it.

- After the last `case` command there is an optional `otherwise` command followed by a group of commands.

- The last line must be an `end` statement.

How does the switch-case statement work?

The value of the switch expression in the `switch` command is compared with the values that are next to each of the `case` statements. If a match is found, the group of commands that follow the `case` statement with the match are executed. (Only one group of commands, the one between the `case` that matches and either the `case`, `otherwise` or `end` statement that is next, is executed).

- If there is more than one match, only the first matching case is executed.

- If no match is found and the `otherwise` statement (which is optional) exists, the group of commands between `otherwise` and `end` is executed.

- If no match is found and the `otherwise` statement does not exist, none of the command groups is executed.

- A case statement can have more than one value. This is done by typing the values in the form: `{value1, value2, value3,...}`. (This form, which is not covered in this book, is called a cell array). The case is executed if at least one of the values matches the value of switch expression.

A NOTE: In MATLAB only the first matching case is executed. After the group of commands that are associated with the first matching are executed, the program skips to the `end` statement. This is different than the C language where break statements are required.

Sample Problem 7-4: Converting units of energy

Write a program in a script file that converts a quantity of energy (work) given in units of either Joule, ft-lb, cal, or eV to the equivalent quantity in different units specified by the user. The program asks the user to enter the quantity of energy, its current units, and the new desired units. The output is the quantity of energy in the new units.

The conversion factors are: $1 \text{ J} = 0.738 \text{ ft-lb} = 0.239 \text{ cal} = 6.24 \times 10^{18} \text{ eV}$.
Use the program to:
a)   Convert 150 J to ft-lb.
b)   2800 cal to Joules.
c)   2.7 eV to cal.

Solution

The program includes two sets of `switch-case` statements and one `if-else-end` statement. The first `switch-case` statement is used to convert the input quantity from its initial units to units of Joules. The second is used to convert the quantity from Joules to the specified new units. The `if-else-end` statement is used to generate an error message if units are entered incorrectly.

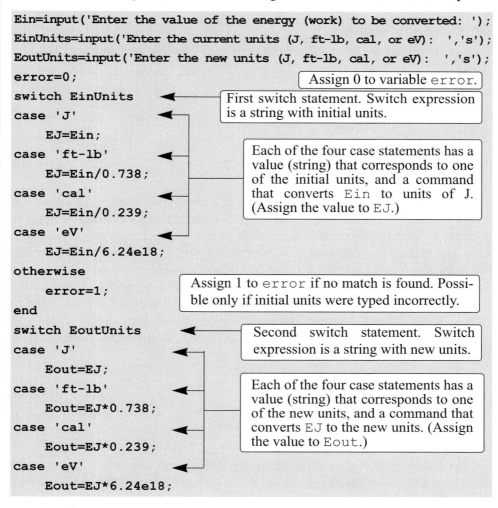

```
Ein=input('Enter the value of the energy (work) to be converted: ');
EinUnits=input('Enter the current units (J, ft-lb, cal, or eV): ','s');
EoutUnits=input('Enter the new units (J, ft-lb, cal, or eV): ','s');
error=0;
switch EinUnits
case 'J'
    EJ=Ein;
case 'ft-lb'
    EJ=Ein/0.738;
case 'cal'
    EJ=Ein/0.239;
case 'eV'
    EJ=Ein/6.24e18;
otherwise
    error=1;
end
switch EoutUnits
case 'J'
    Eout=EJ;
case 'ft-lb'
    Eout=EJ*0.738;
case 'cal'
    Eout=EJ*0.239;
case 'eV'
    Eout=EJ*6.24e18;
```

Assign 0 to variable `error`.

First switch statement. Switch expression is a string with initial units.

Each of the four case statements has a value (string) that corresponds to one of the initial units, and a command that converts `Ein` to units of J. (Assign the value to `EJ`.)

Assign 1 to `error` if no match is found. Possible only if initial units were typed incorrectly.

Second switch statement. Switch expression is a string with new units.

Each of the four case statements has a value (string) that corresponds to one of the new units, and a command that converts `EJ` to the new units. (Assign the value to `Eout`.)

```
otherwise
    error=1;
```
> Assign 1 to error if no match is found. Possible only if new units were typed incorrectly.

```
end
if error
```
> If-else-if statement.

```
    disp('ERROR current or new units are typed incorrectly.')
else
```
> If error is true (nonzero), display an error message.

```
    fprintf('E = %g %s',Eout,EoutUnits)
```

```
end
```
> If error is false (zero), display converted energy.

As an example, the script file (saved as EnergyConversion) is used next in the Command Window to make the conversion in part *b* of the problem statement.

```
>> EnergyConversion
Enter the value of the energy (work) to be converted: 2800
Enter the current units (J, ft-lb, cal, or eV):  cal
Enter the new units (J, ft-lb, cal, or eV):  J
E = 11715.5 J
```

7.4 LOOPS

A loop is another method to alter the flow of a computer program. In a loop, the execution of a command, or a group of commands, is repeated several times consecutively. Each round of execution is called a pass. In each pass at least one variable, but usually more than one, or even all the variables that are defined within the loop are assigned new values. MATLAB has two kinds of loops. In for-end loops (Section 7.4.1) the number of passes is specified when the loop starts. In while-end loops (Section 7.4.2) the number of passes is not known ahead of time, and the looping process continues until a specified condition is satisfied. Both kinds of loops can be terminated at any time with the break command (see Section 7.6).

7.4.1 for-end *Loops*

In for-end loops the execution of a command, or a group of commands, is repeated a predetermined number of times. The form of the loop is shown in Figure 7-5.

- The loop index variable can have any variable name (usually i, j, k, m, and n are used). i and j should not be used if MATLAB is used with complex numbers).

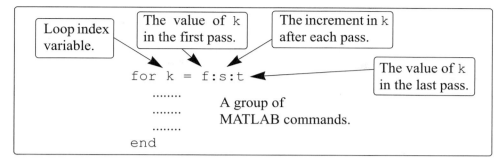

Figure 7-5: The structure of a `for-end` **loop.**

- In the first pass k = f and the computer executes the commands between the `for` and the `end` commands. Then, the program goes back to the `for` command for the second pass. k obtains a new value equal to k = f + s, and the commands between the `for` and the `end` commands are executed with the new value of k. The process repeats itself until the last pass where k = t. Then the program does not go back to the `for`, but continues with the commands that follow the `end` command. For example, if k = 1:2:9, there are five loops, and the value of k in each loop is 1, 3, 5, 7, and 9.

- The increment s can be negative (i.e. k = 25:–5:10 produces four passes with k = 25, 20, 15, 10).

- If the increment value s is omitted, the value is 1 (default) (i.e. k = 3:7 produces five passes with k = 3, 4, 5, 6, 7).

- If f = t, the loop is executed once.

- If f > t and s > 0, or if f < t and s < 0 the loop is not executed.

- If the values of k, s, and t are such that k cannot be equal to t, then, if s is positive, the last pass is the one where k has the largest value that is smaller than t (i.e. k = 8:10:50 produces five passes with k = 8, 18, 28, 38, 48). If s is negative the last pass is the one where k has the smallest value that is larger than t.

- In the `for` command k can also be assigned specific value (typed in as a vector). For example: for k = [7 9 –1 3 3 5].

- The value of k should not be redefined within the loop.

- Each `for` command in a program **must** have an `end` command.

- The value of the loop index variable (k) is not displayed automatically. It is possible to display the value in each pass (sometimes useful for debugging) by typing k as one of the commands in the loop.

- When the loop ends, the loop index variable (k) has the value that was last assigned to it.

A simple example of a for-end loop (in a script file) is:

```
for k=1:3:10
        x = k^2
end
```

When this program is executed, the loop is executed four times. The value of k in the four passes is k = 1, 4, 7, and 10, which means that the values that are assigned to x in each of the passes are x = 1, 16, 49, and 100, respectively. Since a semicolon is not typed at the end of the second line, the value of x is displayed in the Command Window at each pass. When the script file is executed, the display in the Command Window is:

```
>> x =
        1
x =
        16
x =
        49
x =
        100
```

Sample Problem 7-5:   Sum of series

a) Use a for-end loop in a script file to calculate the sum of the first n terms of the series: $\sum_{k=1}^{n} \frac{(-1)^k k}{2^k}$. Execute the script file for $n = 4$ and $n = 20$.

b) The function sin(x) can be written as a Taylor series by:

$$\sin x = \sum_{k=0}^{\infty} \frac{(-1)^k x^{2k+1}}{(2k+1)!}$$

Write a user-defined function file that calculates sin(x) by using the Taylor's series. For the function name and arguments use y = Tsin(x,n). The input arguments are the angle x in degrees, and n the number of terms in the series. Use the function to calculate sin($150°$) using 3 and 7 terms.

Solution

a)   A script file that calculates the sum of the first n terms of the series is shown below.

```
n=input('Enter the number of terms ' );
S=0;                    Setting the sum to zero.
for k=1:n                                    In each pass one element of the
    S=S+(-1)^k*k/2^k;   for-end              series is calculated and is added
                        loop.                to the sum of the elements from
end                                          the previous passes.
fprintf('The sum of the series is: %f',S)
```

The summation is done with a loop. In each pass one term of the series is calculated (in the first pass the first term, in the second pass the second term, and so on), and is added to the sum of the previous elements. The file is saved as Exp7-4a and then executed twice in the Command Window:

```
>> Exp7_4a
Enter the number of terms 4
The sum of the series is: -0.125000
>> Exp7_4a
Enter the number of terms 20
The sum of the series is: -0.222216
```

b)   A user-defined function file that calculates $\sin(x)$ by adding n terms of Taylor's formula is shown below.

```
function y = Tsin(x,n)
% Tsin calculates the sin using Taylor formula.
% Input arguments:
% x The angle in degrees, n number of terms.

xr=x*pi/180;             Converting the angle from degrees to radians.
y=0;
for k=0:n-1
    y=y+(-1)^k*xr^(2*k+1)/factorial(2*k+1);   for-end
                                               loop.
end
```

The first element corresponds to $k = 0$ which means that in order to add n terms of the series, in the last loop $k = n - 1$. The function is used in the Command Window to calculate the $\sin(150^\circ)$ using 3 and 7 terms:

```
>> Tsin(150,3)          Calculating sin(150°) with 3 terms of Taylor series.
ans =
      0.6523
```

```
>> Tsin(150,7)          Calculating sin(150°) with 7 terms of Taylor series.

ans =

      0.5000                    The exact value is 0.5.
```

A note about `for-end` **loops and element-by-element operations:**

In some situations the same end result can be obtained by either using `for-end` loops or by using element-by-element operations. Sample Problem 7-5 illustrates how the `for-end` loop works, but the problem can also be solved by using element-by-element operations (see Problems 7 and 8 in Section 3.9). Element-by-element operations with arrays are one of the superior features of MATLAB that provides the means for computing in circumstances that otherwise require loops. In general, element-by-element operations are faster than loops and are recommended when either method can be used.

Sample Problem 7-6:   Modify vector elements

A vector is given by: $V = [5, 17, -3, 8, 0, -1, 12, 15, 20, -6, 6, 4, -7, 16]$. Write a program as a script file that doubles the elements that are positive and are divisible by 3 and/or 5, and raise to the power of 3 the elements that are negative but greater than -5.

Solution

The problem is solved by using a `for-end` loop that has an `if-elseif-end` conditional statement inside. The number of passes is equal to the number of elements in the vector. In each pass one element is checked by the conditional statement. The element is changed if it satisfies the conditions in the problem statement. A program in a script file that carries out the required operations is:

```
V=[5, 17, -3, 8, 0, -7, 12, 15, 20 -6, 6, 4, -2, 16];
n=length(V);          Setting n to be equal to the number of elements in V.
for k=1:n
    if V(k)>0 & (rem(V(k),3)==0 | rem(V(k),5)==0)      for-end
        V(k)=2*V(k);                                   loop.
    elseif V(k) < 0 & V(k) > -5
        V(k)=V(k)^3;                                   if-
    end                                                elseif-
end                                                    end
V                                                      statement.
```

The program is executed in the Command Window:

```
>> Exp7_5
V =
    10    17   -27     8     0    -7    24    30    40    -6    12     4
  -8    16
```

7.4.2 while-end *Loops*

while-end loops are used in situations when looping is needed but the number of passes is not known ahead of time. In while-end loops the number of passes is not specified when the looping process starts. Instead, the looping process continues until a stated condition is satisfied. The structure of a while-end loop is shown in Figure 7-6.

```
while  conditional expression
    ........
    ........        A group of
    ........        MATLAB commands.
end
```

Figure 7-6: The structure of a while-end **loop.**

The first line is a while statement that includes a conditional expression. When the program reaches this line the conditional expression is checked. If it is false (0), MATLAB skips to the end statement and continues with the program. If the conditional expression is true (1), MATLAB executes the group of commands that follow between the while and end command. Then MATLAB jumps back to the while command and checks the conditional expression. This looping process continues until the conditional expression is false.

For a while-end **loop to execute properly:**

- The conditional expression in the while command must include at least one variable.

- The variables in the conditional expression must have assigned values when MATLAB executes the while command for the first time.

- At least one of the variables in the conditional expression must be assigned a new value in the commands that are between the while and the end. Otherwise, once the looping starts it will never stop since the conditional expression will remain true.

An example of a simple while-end loop is shown in the following program. In this program a variable x with an initial value of 1 is doubled in each pass as long as its value is equal to or smaller than 15.

```
x=1                                          Initial value of x is 1.
while x<=15                 The next command is executed only if x <= 15.
    x=2*x                                    In each pass x doubles.
end
```

When this program is executed the display in the Command Window is:

```
x =                                          Initial value of x.
    1

x =
    2

x =                                          In each pass x doubles.
    4

x =
    8

x =                          When x = 16, the conditional expression in the
    16                       while command is false and the looping stops.
```

Important note:

When writing a `while-end` loop, the programmer has to be sure that the variable (or variables) that are in the conditional expression and are assigned new values during the looping process will eventually be assigned values that make the conditional expression in the `while` command false. Otherwise the looping will continue indefinitely (indefinite loop). For example, in the example above if the conditional expression is changed to be $x >= 0.5$, the looping will continue indefinitely. Such a situation can be avoided by counting the passes and stopping the looping if the number of passes exceeds some large value. This can be done by adding the maximum number of passes to the conditional expression, or by using the break command (Section 7.6).

Since no one is free from making mistakes, a situation of indefinite loop can occur in spite of careful programming. If this happens, the user can stop the execution of an indefinite loop by pressing the **Ctrl + C** or **Ctrl + Break** keys.

Sample Problem 7-7: Taylor series representation of a function

The function $f(x) = e^x$ can be represented in a Taylor series by $e^x = \sum_{n=0}^{\infty} \frac{x^n}{n!}$.

Write a program in a script file that determines e^x by using the Taylor series representation. The program calculates e^x by adding terms of the series and stopping when the absolute value of the term that was added last is smaller than 0.0001. Use a `while-end` loop, but limit the number of passes to 30. If in the 30th pass the value of the term that is added is not smaller than 0.0001, the program stops and displays a message that more than 30 terms are needed.

Use the program to calculate e^2, e^{-4}, and e^{21}.

Solution

The first few terms of the Taylor series are:

$$e^x = 1 + x + \frac{x^2}{2!} + \frac{x^3}{3!} + \dots$$

A program that uses the series to calculate the function is shown next. The program asks the user to enter the value of x. Then, the first term `an`, is assigned the number 1, and `an` is assigned to the sum `S`. Then, from the second term and on, the program uses a `while` loop to calculate the nth term of the series and add it to the sum. The program also counts the number of terms n. The conditional expression in the `while` command is true as long as the absolute value of the nth term `an` is larger than 0.0001, and the number of passes n is smaller than 30. This means that if the 30th term is not smaller than 0.0001, the looping stops.

```
x=input('Enter x ' );
n=1; an=1; S=an;
while abs(an) >= 0.0001 & n <= 30          Start of the while loop.
    an=x^n/factorial(n);                   Calculating the nth term.
    S=S+an;                                Adding the nth term to the sum.
    n=n+1;                                 Counting the number of passes.
end                                        End of the while loop.
if n >= 30                                 if-else-end loop.
    disp('More than 30 terms are needed')
else
fprintf('exp(%f) = %f',x,S)
fprintf('\nThe number of terms used is: %i',n)
end
```

The program uses an `if-else-end` statement to display the results. If the looping stopped because the 30th term is not smaller than 0.0001, it displays a message indicating this. If the value of the function is calculated successfully, it displays the value of the function and the number of terms used. When the program executes, the number of passes depends on the value of x. The program (saved as expox) is used to calculate e^2, e^{-4}, and e^{21}:

```
>> expox
Enter x 2                                          Calculating exp(2).
exp(2.000000) = 7.389046
The number of terms used is: 12                    12 terms used.
>> expox
Enter x -4                                          Calculating exp(-4).
exp(-4.000000) = 0.018307
The number of terms used is: 18                     18 terms used.
>> expox
Enter x 21                                     Trying to calculate exp(21).
More than 30 terms are needed
```

7.5 NESTED LOOPS AND NESTED CONDITIONAL STATEMENTS

Loops and conditional statements can be nested within themselves and each other. This means that a loop and/or a conditional statement can start (and end) within another loop and/or conditional statement. There is no limit to the number of loops and conditional statements that can be nested. It must be remembered, however, that each if, case, for, and while statement must have a corresponding end statement. Figure 7-7 shows the structure of a nested for-end loop within

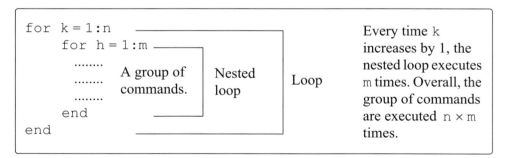

Figure 7-7: Structure of nested loops.

another for-end loop. In the loops shown in this figure, if, for example, n = 3 and m = 4, then first k = 1 and the nested loop executes four times with h = 1, 2, 3, and 4. Next k = 2 and the nested loop executes again four times with h = 1, 2, 3, and 4. Lastly k = 3 and the nested loop executes again four times. Every time a nested loop is typed MATLAB automatically indents the new loop relative to the outside loop. Nested loops and conditional statements are demonstrated in the following sample problem.

Sample Problem 7-8:   Creating a matrix with a loop

Write a program in a script file that creates a $n \times m$ matrix with elements that have the following values. The value of the elements in the first row is the number of the column. The value of the element in the first column is the number of the row. The rest of the elements are equal to the sum of the element above them and the element to the left. When executed, the program asks the user to enter values for n and m.

Solution

The program, shown below, has two loops (one nested), and a nested `if-elseif-else-end` statement. The elements in the matrix are assigned values row by row. The loop index variable of the first loop, `k`, is the address of the row, and the loop index variable of the second loop, `h`, is the address of the column.

```
n=input('Enter the number of rows ');
m=input('Enter the number of columns ');
A=[];                                    Define an empty matrix A
for k=1:n                                Start of the first for-end loop.
    for h=1:m                            Start of the second for-end loop.
        if k==1                          Start of the conditional statement.
            A(k,h)=h;                    Assign values to the elements of the first row.
        elseif h==1
            A(k,h)=k;                    Assign values to the elements of the first column.
        else
            A(k,h)=A(k,h-1)+A(k-1,h);    Assign values to other elements.
        end                              end of the if statement.
    end                                  end of the nested for-end loop.
end                                      end of the first for-end loop.
A
```

The program is executed in the Command Window to create a 4×5 matrix.

```
>> Chap7_exp7
Enter the number of rows 4
Enter the number of columns 5
```

```
A =
     1     2     3     4     5
     2     4     7    11    16
     3     7    14    25    41
     4    11    25    50    91
```

7.6 *THE* break *AND* continue *COMMANDS*

The break **command:**

- When inside a loop (for and while), the break command terminates the execution of the loop (the whole loop, not just the last pass). When the break command appears in a loop, MATLAB jumps to the end command of the loop and continues with the next command (does not go back to the for command of that loop).

- If the break command is inside a nested loop, only the nested loop is terminated.

- When a break command appears outside a loop in a script, or function file, it terminates the execution of the file.

- The break command is usually used within a conditional statement. In loops it provides a method to terminate the looping process if some condition is met. For example, if the number of loops exceeds a predetermined value, or an error in some numerical procedure is smaller than a predetermined value. When typed outside a loop, the break command provides a means to terminate the execution of a file, such as if data transferred into a function file is not consistent with what is expected.

The continue **command:**

- The continue command can be used inside a loop (for and while) to stop the present pass and start the next pass in the looping process.

- The continue command is usually a part of a conditional statement. When MATLAB reaches the continue command, it does not execute the remaining commands in the loop, but skips to the end command of the loop and then starts a new pass.

7.7 EXAMPLES OF MATLAB APPLICATIONS

Sample Problem 7-9:   Withdrawing from a retirement account.

A person in retirement is depositing $300,000 in a saving account which pays 5% interest per year. The person plans to withdraw money from the account once a year. He starts by withdrawing $25,000 after the first year, and then in future years, he increases the amount he withdraws according to the inflation rate. For example, if the inflation rate is 3%, he withdraws $25,750 after the second year. Calculate the number of years that the money in the account will last assuming a constant yearly inflation rate of 2%. Make a plot that shows the yearly withdrawal and the balance of the account over the years.

Solution

The problem is solved by using a loop (a `while` loop since the number of passes in not known before the loop starts). In each pass the amount to be withdrawn and account balance are calculated. The looping continues as long as the account balance is larger than or equal to the amount to be withdrawn. The following is a program in a script file that solves the problem. In the program, `year` is a vector in which each element is a year number, `W` is a vector with the amount withdrawn each year, and `AB` is a vector with the account balance each year.

```
rate=0.05; inf=0.02;
clear W AB year
year(1)=0;                      First element is year 0.
W(1)=0;                         Initial withdrawal amount.
AB(1)=300000;                   Initial account balance.
Wnext=25000;                    The amount to be withdrawn after a year.
ABnext=300000*(1 + rate);       The account balance after a year.
n=2;
    while ABnext >= Wnext       while checks if the next balance
        year(n)=n-1;            is larger then the next withdrawal.
        W(n)=Wnext;             Amount withdrawn in year n – 1.
        AB(n)=ABnext-W(n);      Account balance in year n – 1 after withdrawal.
        ABnext=AB(n)*(1+rate);  The account balance after additional year.
        Wnext=W(n)*(1+inf);
        n=n+1;                  The amount to be withdrawn
    end                         after an additional year.
fprintf('The money will last for %f years',year(n-1))
bar(year,[AB' W'],2.0)
```

The program is executed in the Command Window:

```
>> Chap7_exp9
The money will last for 15 years.
```

The program also generates the following figure (axes labels and legend were added to the plot by using the plot editor).

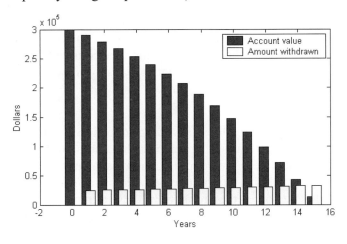

Sample Problem 7-10: Choosing numbers for a lottery

In a lottery the player has to select several numbers out of a list. Write a user-defined function that generates a list of *n* integer numbers that are uniformly distributed between the numbers *a* and *b*. All the selected numbers on the list must be different.

a) Use the function to generate a list of 6 numbers from the numbers 1 through 49.

b) Use the function to generate a list of 8 numbers from the numbers 60 through 75.

c) Use the function to generate a list of 9 numbers from the numbers −15 through 15.

Solution

The function, shown below, uses the MATLAB `rand` function (refer back to Section 3.7). To make sure all the numbers are different from each other, the numbers are picked one by one. Each number that is picked (by the `rand` function) is compared with all the numbers that were picked before. If a match is found, the number is not selected, and a new number is picked.

```
function x = lotto(a,b,n)
% lotto selects n numbers (all different) from the domain a,b.
```

```
% x is a vector with the n numbers.
   x(1)=round((b-a)*rand+a);    Select the first integer between a and b.
for p=2:n
    x(p)=round((b-a)*rand+a);       Assign the next number to x(p).
    r=0;                                         Set r to zero.
    while r==0                             See explanation below.
      r=1;                                          Set r to 1.
      for k=1:p-1    for loop compares the number assigned to x(p) with
                     the numbers that have already been selected.
        if x(k)==x(p)                    If a match if found, a
          x(p)=round((b-a)*rand+a);      new number is
                                         assigned to x(p) and
          r=0;                           r is set to zero.
          break   ◄────      The nested for loop is stopped. The pro-
        end                  gram goes back to the while loop. Since
      end                    r = 0 the nested loop inside the while
    end                      loop starts again and checks if the new
end                          number that is assigned to x(p) is equal
                             to a number that is already in the vector x.
```

The while loop checks that every new number (element) that is added to the vector x does not match any of the elements (previous numbers) that are already in the vector x. If a match is found, it keeps choosing new numbers until the new number is different than all the elements that are already in x.

The function is used next in the Command Window for the three cases stated in the problem statement.

```
>> lotto(1,49,6)

ans =
    23     2    40    22    31    39
>> lotto(60,75,8)

ans =
    67    66    73    68    63    70    60    72
>> lotto(-15,15,9)

ans =
   -13    -4     0    -2     2     3   -12    12     8
```

Sample Problem 7-11: Flight of a model rocket

The flight of a model rocket can be modeled as follows. During the first $0.15\,$s the rocket is propelled up by the rocket engine with a force of $16\,$N. The rocket then flies up while slowing down under the force of gravity. After it reaches the apex, the rocket starts to fall back down. When its down velocity reaches 20 m/s a parachute opens (assumed to open instantly) and the rocket continues to move down at a constant speed of 20 m/s until it hits the ground. Write a program that calculates and plots the speed and altitude of the rocket as a function of time during the flight.

Solution

The rocket is assumed to be a particle that moves along a straight line in the vertical plane. For motion with constant acceleration along a straight line, the velocity and position as a function of time are given by:

$$v(t) \,=\, v_0 + at \quad \text{and} \quad s(t) \,=\, s_0 + v_0 t + \frac{1}{2}at$$

where v_0 and s_0 are the initial velocity and position, respectively. In the computer program the flight of the rocket is divided into three segments. Each segment is calculated in a `while` loop. In every pass the time increases by an increment.

Segment 1: The first $0.15\,$s when the rocket engine is on. During this period, the rocket moves up with a constant acceleration. The acceleration is determined by drawing a free body and a mass acceleration diagram (shown on the right). From Newton's second law, the sum of the forces in the vertical direction is equal to the mass times the acceleration (equilibrium equation):

$$+\!\uparrow \Sigma F \,=\, F_E - mg \,=\, ma$$

Solving the equation for the acceleration gives:

$$a \,=\, \frac{F_E - mg}{m}$$

The velocity and height as a function of time are:

$$v(t) \,=\, 0 + at \quad \text{and} \quad h(t) \,=\, 0 + 0 + \frac{1}{2}at^2$$

where the initial velocity and initial position are both zero. In the computer program this segment starts when $t = 0$, and the looping continues as long as $t < 0.15$ s. The time, velocity, and height at the end of this segment are t_1, v_1, and h_1.

Segment 2: The motion from when the engine stops until the parachute opens. In this segment the rocket moves with a constant deceleration g. The speed and height of the rocket as a function of time are given by:

$$v(t) = v_1 - g(t - t_1) \quad \text{and} \quad h(t) = h_1 + v_1(t - t_1) - \frac{1}{2}g(t - t_1)^2$$

In this segment the looping continues until the velocity of the rocket is $-20\,\text{m/s}$ (negative since the rocket moves down). The time and height at the end of this segment are t_2, and h_2.

Segment 3: The motion from when the parachute opens until the rocket hits the ground. In this segment the rocket moves with constant velocity (zero acceleration). The height as a function of time is given by: $h(t) = h_2 - v_{chute}(t - t_2)$, where v_{chute} is the constant velocity after the parachute opens. In this segment the looping continues as long as the height is greater than zero.

A program in a script file that carries out the calculations is shown below:

```
m=0.05; g=9.81; tEngine=0.15; Force=16; vChute=-20; Dt=0.01;
clear t v h
n=1;
t(n)=0; v(n)=0; h(n)=0;
% Segment 1
a1=(Force-m*g)/m;
while t(n) < tEngine & n < 50000          The first while loop.
    n=n+1;
    t(n)=t(n-1)+Dt;
    v(n)=a1*t(n);
    h(n)=0.5*a1*t(n)^2;
end
v1=v(n); h1=h(n); t1=t(n);
% Segment 2
while v(n) >= vChute & n < 50000          The second while loop.
    n=n+1;
    t(n)=t(n-1)+Dt;
    v(n)=v1-g*(t(n)-t1);
    h(n)=h1+v1*(t(n)-t1)-0.5*g*(t(n)-t1)^2;
end
v2=v(n); h2=h(n); t2=t(n);
% Segment 3
while h(n) > 0 & n < 50000                The third while loop.
```

```
    n=n+1;
    t(n)=t(n-1)+Dt;
    v(n)=vChute;
    h(n)=h2+vChute*(t(n)-t2);
end
subplot(1,2,1)
plot(t,h,t2,h2,'o')
subplot(1,2,2)
plot(t,v,t2,v2,'o')
```

The accuracy of the results depends on the magnitude of the time increment Dt. An increment of 0.01s appears to give good results. The conditional expression in the while commands also includes a condition for n (if n is larger than 50,000 the loop stops). This is done as a precaution to avoid an infinite loop in case there is an error in the statements inside the loop. The plots generated by the program are shown below (axes labels and text were added to the plots using the plot editor).

Note:

The problem can be solved and programmed in different ways. The solution shown here is one option. For example, instead of using while loops, the times that the parachute opens and the rocket hits the ground can be calculated first, and then for-end loops can be used instead of the while loop. If the times are determined first, it is possible also to use element-by-element calculations instead of loops.

Sample Problem 7-12: AC to DC convertor

A half-wave diode rectifier is an electrical circuit that converts AC voltage to DC voltage. A rectifier circuit that consists of an AC voltage source, a diode, a capacitor, and a load (resistor) is shown in the figure. The voltage of the source is $v_s = v_0 \sin(\omega t)$,

where $\omega = 2\pi f$, in which f is the frequency. The operation of the circuit is illustrated in the diagram on the right where the dashed line shows the source voltage, and the solid line shows the voltage across the resistor. In the first cycle, the diode is on (conducting current) from $t = 0$ until

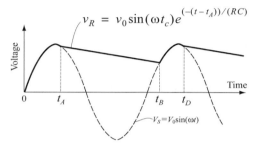

$t = t_A$. At this time the diode turns off and the power to the resistor is supplied by the discharging capacitor. At $t = t_B$ the diode turns on again and continues to conduct current until $t = t_D$. The cycle continues as long as the voltage source is on. In a simplified analysis of this circuit, the diode is assumed to be ideal and the capacitor is assumed to have no charge initially (at $t = 0$). When the diode is on, the resistor's voltage and current are given by:

$$v_R = v_0 \sin(\omega t) \quad \text{and} \quad i_R = v_0 \sin(\omega t)/R$$

The current in the capacitor is:

$$i_C = \omega C v_0 \cos(\omega t)$$

When the diode is off, the voltage across the resistor is given by:

$$v_R = v_0 \sin(\omega t_A) e^{(-(t - t_A))/(RC)}$$

The times when the diode switches off (t_A, t_D, and so on) are calculated from the condition $i_R = -i_C$. The diode switches on again when the voltage of the source reaches the voltage across the resistor (time t_B in the figure).

Write a MATLAB program that plots the voltage across the resistor v_R and the voltage of the source v_s as a function of time for $0 \le t \le 70$ ms. The resistance of the load is 1800 Ω, the voltage source $v_0 = 12$ V, and $f = 60$ Hz. To examine the effect of the capacitor size on the voltage across the load, execute the program twice; once with $C = 45$ μF, and once with $C = 10$ μF.

Solution

A program that solves the problem is listed below. The program has two parts. One that calculates the voltage v_R when the diode is on, and the other when the diode is off. The `switch` command is used for switching between the two parts. The calculations start with the diode on (the variable `state='on'`), and when $i_R - i_C \leq 0$ the value of `state` is changed to `'off'`, and the program switches to the commands that calculate v_R for this state. These calculations continue until $v_s \geq v_R$ when the program switches back to the equations that are valid when the diode is on.

```
V0=12; C=45e-6; R=1800; f=60;
Tf=70e-3; w=2*pi*f;
clear t VR Vs
t=0:0.05e-3:Tf;
n=length(t);
state='on'                          Assign 'on' to the variable state.
for i=1:n
    Vs(i)=V0*sin(w*t(i));           Calculate the voltage of the source at time t.
    switch state
        case 'on'                   Diode is on.
        VR(i)=Vs(i);
        iR=Vs(i)/R;
        iC=w*C*V0*cos(w*t(i));
        sumI=iR+iC;
        if sumI <= 0                Check if iR - iC ≤ 0.
            state='off';            If true, assign 'off' to state.
            tA=t(i);                Assign value to tA.
        end
        case 'off'                  Diode is off.
        VR(i)=V0*sin(w*tA)*exp(-(t(i)-tA)/(R*C));
        if Vs(i) >= VR(i)           Check if vs ≥ vR.
            state='on';             If true, assign
        end                         'on' to the
    end                             variable state.
end
plot(t,Vs,':',t,VR,'k','linewidth',1)
xlabel('Time (s)'); ylabel('Voltage (V)')
```

The two plots generated by the program are shown below. One plot shows the result with $C = 45$ μF, and the other with $C = 10$ μF. It can be observed that with larger capacitor the DC voltage is smoother (smaller ripple in the wave).

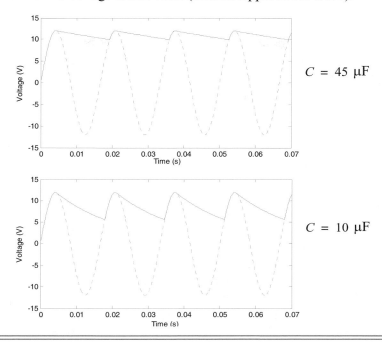

7.8 PROBLEMS

1. Evaluate the following expressions without using MATLAB. Check the answer with MATLAB.

   a) $14 > 15/3$

   b) $y = 8/2 < 5 \times 3 + 1 > 9$

   c) $y = 8/(2 < 5) \times 3 + (1 > 9)$

   d) $2 + 4 \times 3 \sim = 60/4 - 1$

2. Given: $a = 4$, $b = 7$. Evaluate the following expressions without using MATLAB. Check the answer with MATLAB.

   a) $y = a + b >= a \times b$

   b) $y = a + (b > = a) \times b$

   c) $y = b - a < a < a/b$

3. Given: $v = [4 \ -2 \ -1 \ 5 \ 0 \ 1 \ -3 \ 8 \ 2]$ and $w = [0 \ 2 \ 1 \ -1 \ 0 \ -2 \ 4 \ 3 \ 2]$. Evaluate the following expressions without using MATLAB. Check the answer with MATLAB.

   a) $v <= w$ b) $w = v$

   c) $v < w + v$ d) $(v < w) + v$

4. Use the vectors *v* and *w* from the previous problem. Use relational operators to create a vector *y* that is made up from the elements of *w* that are smaller than the elements of *v*.

5. Evaluate the following expressions without using MATLAB. Check the answer with MATLAB.
   a) -3&0   *b*) 4<-1&5>0
   c) 8-12|6+5&~-2   *d*) ~4&0+8*~(4|0)

6. The maximum daily temperature (in °F) for New York City and Anchorage, Alaska during the month of January, 2001 are given in the vectors below (data from the U.S. National Oceanic and Atmospheric Administration).
   TNY = [31 26 30 33 33 39 41 41 34 33 45 42 36 39 37 45 43 36 41 37 32 32 35 42 38 33 40 37 36 51 50]
   TANC = [37 24 28 25 21 28 46 37 36 20 24 31 34 40 43 36 34 41 42 35 38 36 35 33 42 42 37 26 20 25 31]
   Write a program in a script file to answer the following:
   a) Calculate the average temperature for the month in each city.
   b) How many days was the temperature below the average in each city?
   c) How many days, and which dates in the month, was the temperature in Anchorage higher then the temperature in New York?
   d) How many days, and which dates in the month, was the temperature the same in both cities?
   e) How many days, and which dates in the month, was the temperature in both cities above freezing (above 32 °F)?

7. Use MATLAB in the two different ways described below to plot the function:
$$f(x) = \begin{cases} 20 & \text{for} & x \le -1 \\ -5x + 10 & \text{for} & -1 \le x \le 1 \\ -10x^2 + 35x - 20 & \text{for} & 1 \le x \le 3 \\ -5x + 10 & \text{for} & 3 \le x \le 4 \\ -10 & \text{for} & x \ge 4 \end{cases}$$
   in the domain $-2 \le x \le 5$.

   a) Write a program in a script file, using conditional statements and loops.
   b) Create a user-defined function for $f(x)$, and then use the function in a script file to make the plot.

8. Use loops to create a 3×5 matrix in which the value of each element is the difference between the indices divided by the sum of its indices (the row number and column number of the element). For example, the value of the element (2,5) is $(2-5)/(2+5) = -0.4286$.

9. Write a program in a script file that determines the real roots of a quadratic equation $ax^2 + bx + c = 0$. Name the file `quadroots`. When the file runs it asks the user to enter the values of the constants a, b, and c. To calculate the roots of the equation the program calculates the discriminant D given by:

$$D = b^2 - 4ac$$

If $D > 0$ the program displays a message: "The equation has two roots," and the roots are displayed in the next line.
If $D = 0$ the program displays a message: "The equation has one root," and the root is displayed in the next line.
If $D < 0$ the program displays a message: "The equation has no real roots."
Run the script file in the Command Window three times to obtain solutions to the following three equations:

a) $-2x + 16x - 32 = 0$

b) $8x^2 + 9x + 3 = 0$

c) $3x^2 + 5x - 6 = 0$

10. Write a program (using a loop) that determines the expression:

$$\left(12 \sum_{n=1}^{m} \frac{(-1)^n}{n^2} \right)^{\frac{1}{2}} \quad (n = 1, 2,..., m)$$

Run the program with $m = 10$, $m = 1000$, and $m = 10,000$. Compare the result with π.

11. A vector is given by: $x = [15\ -6\ 0\ 8\ -2\ 5\ 4\ -10\ 0.5\ 3]$. Using conditional statements and loops, write a program that determines and displays the sum of the positive elements of the vector, and the sum of the negative elements of the vector.

12. The geometric mean GM of a set of n positive numbers $x_1, x_2, ..., x_n$ is defined by:

$$GM = (x_1 \cdot x_2 \cdot ... \cdot x_n)^{1/n}$$

Write a user-defined function that calculates the geometric mean of a set of numbers. For function name and arguments use `GM=Geomean(x)`, where the input argument x is a vector of numbers (any length), and the output argument GM is their geometric mean. The average mean is useful for calculating the average return of a stock. The following table gives the return of an IBM stock in the last ten years (a return of 16% means 1.16 etc.). Use the user-defined function `Geomean` to calculate the average return of the stock.

Year	1997	1998	1999	2000	2001	2002	2003	2004	2005	2006
Return	1.38	1.76	1.17	0.79	1.42	0.64	1.2	1.06	0.83	1.18

13. The factorial $n!$ of a positive number (integer) is defined by: $n! = n \cdot (n-1) \cdot (n-2) \cdot \ldots \cdot 3 \cdot 2 \cdot 1$, where $0! = 1$. Write a user-defined function that calculates the factorial $n!$ of a number. For function name and arguments use `y=fact(x)`, where the input argument x is the number whose factorial is calculated, and the output argument y is the value $x!$. The function displays an error message if a negative or a non-integer number is entered when the function is called. Use `fact` with the following numbers: *a*) 12! *b*) 0! *c*) -7! *d*) 6.7!

14. The Taylor's series expansion for $\cos(x)$ is:

$$\cos(x) = 1 - \frac{x^2}{2!} + \frac{x^4}{4!} - \frac{x^6}{6!} + \ldots = \sum_{n=0}^{\infty} \frac{(-1)^n}{(2n)!} x^{(2n)}$$

where x is in radians. Write a user-defined function that determines $\cos(x)$ using Taylor's series expansion. For function name and arguments use `y=cosTaylor(x)`, where the input argument x is the angle in degrees and the output argument y is the value for $\cos(x)$. In the program of the user-defined function use a loop for adding the terms of the Taylor's series. If a_n is the nth term in the series, than the sum S_n of the n terms is $S_n = S_{n-1} + a_n$. In each pass calculate the estimated error E given by: $E = \left| \frac{S_n - S_{n-1}}{S_{n-1}} \right|$. Stop adding terms when $E \le 0.000001$. If you have a solution for Problem 13, insert it as a subfunction inside `cosTaylor`, and use it for calculating the factorial term in the equation (otherwise use MATLAB's built-in function `factorial`).
Use `sinTaylor` for calculating: *a*) $\cos(55°)$ *b*) $\cos(190°)$. Compare the values with the values obtained by using a calculator.

15. Write a program in a script file that finds the smallest even integer that is also divisible by 7 and whose cube is greater than 40,000. Use a loop in the program. The loop should start from 1 and stop when the number is found. The program prints a message: "The required number is:" and then prints the number.

16. Write a user-defined function that finds all the prime numbers between 1 and n. Name the function `pr=prime(n)`, where the input argument n is a positive integer, and the output argument pr is a vector with the prime numbers. If a negative number or a number that is not an integer is entered when the function is called, an error message "The input argument must be a positive integer." is displayed. Use the function with:
a) `prime(30)`.   *b*) `prime(52.5)`.
c) `prime(79)`.   *d*) `prime(-20)`.

17. Write a user-defined function that sorts the elements of a vector (of any length) from the largest to the smallest. For the function name and arguments use y = downsort(x). The input to the function is a vector x of any length, and the output y is a vector in which the elements of x are arranged in a descending order. Do not use the MATLAB built-in functions sort, max, or min. Test your function on a vector with 14 numbers (integers) randomly distributed between –30 and 30. Use the MATLAB rand function to generate the initial vector.

18. Write a user-defined function that sorts the elements of a matrix. For the function name and arguments use B = matrixsort(A), where A is any size matrix and B is a matrix of the same size with the elements of A rearranged in descending order row after row where the (1,1) element is the largest, and the (*m,n*) element is the smallest. Use the user-defined function downsort from Problem 17 as a subfunction within matrixsort.

    Test your function on a 4 × 7 matrix with elements (integers) randomly distributed between –30 and 30. Use the MATLAB rand function to generate the initial matrix.

19. The Karvonen formula is one method for calculating the training heart rate (*THR*).

    $$THR = [(220 - AGE) - RHR] \times INTEN + RHR$$

    where *AGE* is the age, *RHR* the rest heart rate, and *INTEN* is the fitness level (0.55 for low, 0.65 for medium, and 0.8 for high fitness level). Write a program in a script file that determines the *THR*. The program asks the user to enter his/her age (number), rest heart rate (number), and fitness level (low, medium, or high). The program then displays the training heart rate.

20. Write a user-defined function that determines the polar coordinates of a point from the Cartesian coordinates in a two-dimensional plane. For the function name and arguments use [theta radius]=CartesianTo-Polar(x,y). The input arguments are the *x* and *y* coordinates of the point, and the output arguments are the angle θ and the radial distance to the point. The angle θ is in degrees and is measured relative to the positive *x* axis, such that it is a positive number in quadrants I, II, and III, and a negative number in the IV quadrant. Use the function to determine the polar coordinates of points (15, 3), (–7, 12), (–17, –9), and (10, –6.5).

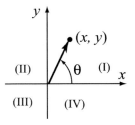

21. Write a program in a script file that calculates the cost of renting a car according to the following price schedule:

Type of car	Rental period		
	1-6 days	**7-27 days**	**28-60 days**
Class B	$27 per day	$162 for 7 days, + $25 for each additional day.	$662 for 28 days, + $23 for each additional day.
Class C	$34 per day	$204 for 7 days, + $31 for each additional day.	$284 for 28 days, + $28 for each additional day.
Class D	Class D cannot be rented for less than 7 days.	$276 for 7 days, + $43 for each additional day.	$1,136 for 28 days, + $38 for each additional day.

The program asks the user to enter the rental period and the type of car. The program then displays the cost. If a period longer than 60 days is entered, a message "Rental is not available for more than 60 days" is displayed. If a rental period of less than 6 days is entered for class D, a message "Class D cars cannot be rented for less than 6 days" is displayed.

Run the program nine times for the following cases:
Class B, for 3, 14, and 50 days. Class C, for 20, 28 and 61 days. Class D for 6, 18, and 60 days.

22. A fuel tank is made of a rectangular prism center and half circular cylinders at the top and bottom as shown, where $r = 20$ cm, $H = 15$ cm, and $L = 60$ cm.

Write a user-defined function (for the function name and arguments use V = Volfuel(h)) that gives the volume of the fuel in the tank (in liters) as a function of the height h (measured from the bottom). Use the function to make a plot of the volume as a function of h for $0 \le h \le 70$ cm.

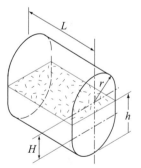

23. The position x as a function of time, of a particle that moves along a straight line is shown in the figure, and is given by the following equations:

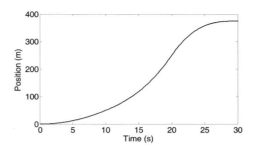

$$x(t) = \begin{cases} 0.5t^2 \text{ m} & \text{for} \quad 0 \le t \le 10 \text{ s} \\ 0.05t^3 - t^2 + 15t - 50 \text{ m} & \text{for} \quad 10 \le t \le 20 \text{ s} \\ 0.0025t^4 - 0.15t^3 + 135t - 1650 \text{ m} & \text{for} \quad 20 \le t \le 30 \text{ s} \end{cases}$$

Write three user-defined functions: one calculates the position of the particle at time t (for the function name and arguments use $v = $ `position(t)`). The second function calculates the velocity of the particle at time t (for the function name and arguments use $v = $ `velocity(t)`), and the third function calculates the acceleration of the particle at time t (for the function name and arguments use $a = $ `acceleration(t)`). In a script file, write a program that creates plots of the position, velocity, and acceleration as function of time (three plots on the same page). In the program, first create a vector t, $0 \le t \le 30$ s, and then use the functions `position`, `velocity`, and `acceleration` to create vectors of velocity and acceleration that are used for the plots.

24. Write a program that determines the change given back to a customer in a self-service checkout machine of a supermarket for purchases of up to $10. The program generates a random number between 0.01 and 10.00 and displays the number as the amount to be paid. The program then asks the user to enter his payment, which can be one $1 bill, one $5 bill, or one $10. If the payment is less than the amount to be paid, an error message is displayed. If the payment is sufficient, the program calculates the change and lists the bills and/or the coins that make up the change, which has to be composed of the least number of bills and coins. For example, if the amount to be paid is $2.33, and a $10 bill is entered as payment, then the change is: 1 $5 bill, 2 $1 bills, 2 quarters, 1 dime, 1 nickel, and 2 cents.

Write the main program in a script file, and write two user-defined functions that are used in the script file. One user-defined function generates a random number between 0.01 and 10.00. The other user-defined function calculates the composition of the change.

25. The concentration of a drug in the body C_P can be modeled by the equation:

$$Cp = \frac{D_G}{V_d} \frac{k_a}{(k_a - k_e)} (e^{-k_e t} - e^{-k_a t})$$

where D_G is the dosage administrated (mg), V_d is the volume of distribution (L), k_a is the absorption rate constant (h^{-1}), k_e is the elimination rate constant (h^{-1}), and t is the time (h) since the drug was administered. For a certain drug, the following quantities are given: $D_G = 150$ mg, $V_d = 50$ L, $k_a = 1.6$ h^{-1}, and $k_e = 0.4$ h^{-1}.

a) A single dose is administered at $t = 0$. Calculate and plot C_P versus t for 10 hours.

a) A first dose is administered at $t = 0$, and subsequently four more doses are administered at intervals of 4 hours (i.e. at $t = 4, 8, 12, 16$). Calculate and plot C_P versus t for 24 hours.

26. The solution of the nonlinear equation $x^3 - P = 0$ gives the cubic root of the number P. A numerical solution of the equation can be calculated with Newton's method. The solution process starts by choosing a value x_1 as a first estimate of the solution. Using this value, a second (more accurate) solution x_2 can be calculated with $x_2 = x_1 - \frac{x_1^3 - P}{3x_1^2}$, which is then used for calculating a third, more accurate, solution x_3, and so on. The general equation for calculating the value of the solution x_{i+1} from the solution x_i is: $x_{i+1} = x_i - \frac{x_i^3 - P}{3x_i^2}$. Write a user-defined function that calculates the cubic root of a number. For function name and arguments use y=cubic(P), where the input argument P is the number whose cubic root is to be determined, and the output argument y is the value $\sqrt[3]{P}$. In the program use $x = P$ for the first estimate of the solution. Then, by using the general equation in a loop, calculate new, more accurate solutions. Stop the looping when the estimated relative error E defined by $E = \left| \frac{x_{i+1} - x_i}{x_i} \right|$ is smaller than 0.00001.

Use the function cubic to calculate:

a) $\sqrt[3]{100}$ b) $\sqrt[3]{9261}$ c) $\sqrt[3]{-70}$

Chapter 8

Polynomials, Curve Fitting, and Interpolation

Polynomials are mathematical expressions that are frequently used for problem solving and modeling in science and engineering. In many cases an equation that is written in the process of solving a problem is a polynomial, and the solution of the problem is the zero of the polynomial. MATLAB has a wide selection of functions that are specifically designed for handling polynomials. How to use polynomials in MATLAB is described in Section 8.1.

Curve fitting is a process of finding a function that can be used to model data. The function does not necessarily pass through any of the points, but models the data with the smallest possible error. There are no limitations to the type of the equations that can be used for curve fitting. Often, however, polynomial, exponential, and power functions are used. In MATLAB curve fitting can be done by writing a program, or by interactively analyzing data that is displayed in the Figure Window. Section 8.2 describes how to use MATLAB programming for curve fitting with polynomials and other functions. Section 8.4 describes the basic fitting interface which is used for interactive curve fitting and interpolation.

Interpolation is the process of estimating values between data points. The simplest interpolation is done by drawing a straight line between the points. In a more sophisticated interpolation, data from additional points is used. How to interpolate with MATLAB is discussed in Sections 8.3 and 8.4.

8.1 POLYNOMIALS

Polynomials are functions that have the form:

$$f(x) = a_n x^n + a_{n-1} x^{n-1} + \ldots + a_1 x + a_0$$

The coefficients $a_n, a_{n-1}, \ldots, a_1, a_0$ are real numbers, and n, which is a nonnega-

tive integer, is the degree, or order, of the polynomial.
Examples of polynomials are:

$$f(x) = 5x^5 + 6x^2 + 7x + 3 \qquad \text{polynomial of degree 5.}$$
$$f(x) = 2x^2 - 4x + 10 \qquad \text{polynomial of degree 2.}$$
$$f(x) = 11x - 5 \qquad \text{polynomial of degree 1.}$$

A constant (e.g. $f(x) = 6$) is a polynomial of degree 0.

In MATLAB, polynomials are represented by a row vector in which the elements are the coefficients $a_n, a_{n-1}, \ldots, a_1, a_0$. The first element is the coefficient of the x with the highest power. The vector has to include all the coefficients, including the ones that are equal to 0. For example:

Polynomial	**MATLAB representation**
$8x + 5$	$p = [8 \ \ 5]$
$2x^2 - 4x + 10$	$d = [2 \ \ -4 \ \ 10]$
$6x^2 - 150$, MATLAB form: $6x^2 + 0x - 150$	$h = [6 \ \ 0 \ \ -150]$
$5x^5 + 6x^2 - 7x$, MATLAB form: $5x^5 + 0x^4 + 0x^3 + 6x^2 - 7x + 0$	$c = [5 \ \ 0 \ \ 0 \ \ 6 \ \ -7 \ \ 0]$

8.1.1 Value of a Polynomial

The value of a polynomial at a point x can be calculated with the function `polyval` that has the form:

| `polyval(p,x)` |

p is a vector with the coefficients of the polynomial.

x is a number, or a variable that has an assigned value, or a computable expression.

x can also be a vector or a matrix. In such a case the polynomial is calculated for each element (element-by-element), and the answer is a vector, or a matrix, with the corresponding values of the polynomial.

Sample Problem 8-1: Calculating polynomials with MATLAB

For the polynomial: $f(x) = x^5 - 12.1x^4 + 40.59x^3 - 17.015x^2 - 71.95x + 35.88$
a)   Calculate $f(9)$.
b)   Plot the polynomial for $-1.5 \le x \le 6.7$.

Solution

The problem is solved in the Command Window.
a)   The coefficients of the polynomials are assigned to vector p. The function

`ployval` is then used to calculate the value at $x = 9$.

```
>> p = [1 -12.1 40.59 -17.015 -71.95 35.88];
>> polyval(p,9)
ans =
  7.2611e+003
```

b) To plot the polynomial, a vector `x` is first defined with elements ranging from -1.5 to 6.7. Then a vector `y` is created with the values of the polynomial for every element of `x`. Finally, a plot of `y` vs. `x` is made.

```
>> x=-1.5:0.1:6.7;
>> y=polyval(p,x);
>> plot(x,y)
```

Calculating the value of the polynomial for each element of the vector `x`.

The plot created by MATLAB is (axes labels were added with the Plot Editor):

8.1.2 Roots of a Polynomial

Roots of a polynomial are the values of the argument for which the value of the polynomial is equal to zero. For example, the roots of the polynomial $f(x) = x^2 - 2x - 3$ are the values of x for which $x^2 - 2x - 3 = 0$, which are $x = -1$, and $x = 3$.

MATLAB has a function, called `roots`, that determines the root, or roots, of a polynomial. The form of the function is:

$$r = \text{roots}(p)$$

`r` is a column vector with the roots of the polynomial.

`p` is a row vector with the coefficients of the polynomial.

For example, the roots of the polynomial in Sample Problem 8-1 can be determined by:

```
>> p= 1 -12.1 40.59 -17.015 -71.95 35.88];
>> r=roots(p)
r =
    6.5000
    4.0000
    2.3000
   -1.2000
    0.5000
```

When the roots are known, the polynomial can actually be written as:

$$f(x) = (x + 1.2)(x - 0.5)(x - 2.3)(x - 4)(x - 6.5)$$

The `roots` command is very useful for finding the roots of a quadratic equation. For example, to find the roots of $f(x) = 4x^2 + 10x - 8$ type:

```
>> roots([4 10 -8])
ans =
   -3.1375
    0.6375
```

When the roots of a polynomial are known, the `poly` command can be used for determining the coefficients of the polynomial. The form of the `poly` command is:

p is a row vector with the coefficients of the polynomial.

r is a vector (row or column) with the roots of the polynomial.

For example, the coefficients of the polynomial in Sample Problem 8-1 can be obtained from the roots of the polynomial (see above) by:

```
>> r=6.5 4 2.3 -1.2 0.5];
>> p=poly(r)
p =
    1.0000   -12.1000    40.5900   -17.0150   -71.9500    35.8800
```

8.1.3 Addition, Multiplication, and Division of Polynomials

Addition:

Two polynomials can be added (or subtracted) by adding the vectors of the coefficients. If the polynomials are not of the same order (which means that the vectors of the coefficients are not of the same length), the shorter vector has to be modified to be of the same length as the longer vector by adding zeros (called padding) in front. For example, the polynomials:

$f_1(x) = 3x^6 + 15x^5 - 10x^3 - 3x^2 + 15x - 40$ and $f_2(x) = 3x^3 - 2x - 6$ can be added by:

```
>> p1=[3 15 0 -10 -3 15 -40];
>> p2=[3 0 -2 -6];
>> p=p1+[0 0 0 p2]
p =
      3     15      0     -7     -3     13    -46
```

Three 0's are added in front of p2, since the order of p1 is 6 and the order of p2 is 3.

Multiplication:

Two polynomials can be multiplied with the MATLAB built-in function conv which has the form:

c = conv(a,b)

c is a vector of the coefficients of the polynomial that is the product of the multiplication.

a and b are the vectors of the coefficients of the polynomials that are being multiplied.

- The two polynomials do not have to be of the same order.

- Multiplication of three or more polynomials is done by using the conv function repeatedly.

For example, multiplication of the polynomials $f_1(x)$ and $f_2(x)$ above gives:

```
>> pm=conv(p1,p2)
pm =
      9     45     -6    -78    -99     65    -54    -12    -10    240
```

which means that the answer is:

$9x^9 + 45x^8 - 6x^7 - 78x^6 - 99x^5 + 65x^4 - 54x^3 - 12x^2 - 10x + 240$

Division:

A polynomial can be divided by another polynomial with the MATLAB built-in function deconv which has the form:

[q,r] = deconv(u,v)

q is a vector with the coefficients of the quotient polynomial.
r is a vector with the coefficients of the remainder polynomial.

u is a vector with the coefficients of the numerator polynomial.
v is a vector with the coefficients of the denominator polynomial.

For example, dividing $2x^3 + 9x^2 + 7x - 6$ by $x + 3$ is done by:

```
>> u=[2 9 7 -6];
>> v=[1 3];
```

```
>> [a b]=deconv(u,v)
a =
     2     3    -2
b =
     0     0     0     0
```

The answer is: $2x^2 + 3x - 2$.

Remainder is zero.

An example of division that gives a remainder is $2x^6 - 13x^5 + 75x^3 + 2x^2 - 60$ divided by $x^2 - 5$:

```
>> w=[2 -13 0 75 2 0 -60];
>> z=[1 0 -5];
>> [g h]=deconv(w,z)
g =
     2   -13    10    10    52
h =
     0     0     0     0     0    50   200
```

The quotient is: $2x^4 - 13x^3 + 10x^2 + 10x + 52$.

The remainder is: $50x + 200$.

The answer is: $2x^4 - 13x^3 + 10x^2 + 10x + 52 + \dfrac{50x + 200}{x^2 - 5}$.

8.1.4 Derivatives of Polynomials

The built-in function `polyder` can be used to calculate the derivative of a single polynomial, a product of two polynomials, and a quotient of two polynomials, as shown in the following three commands.

k = polyder(p)	Derivative of a single polynomial. p is a vector with the coefficients of the polynomial. k is a vector with the coefficients of the polynomial that is the derivative.
k = polyder(a,b)	Derivative of a product of two polynomials. a and b are vectors with the coefficients of the polynomials that are multiplied. k is a vector with the coefficients of the polynomial that is the derivative of the product.
[n d] = polyder(u,v)	Derivative of a quotient of two polynomials. u and v are vectors with the coefficients of the numerator and denominator polynomials. n and d are vectors with the coefficients of the numerator and denominator polynomials in the quotient that is the derivative.

The only difference between the last two commands is in the number of output arguments. With two output arguments MATLAB calculates the derivative of quotient of two polynomials. With one output argument the derivative is of the product.

For example, if $f_1(x) = 3x^2 - 2x + 4$, and $f_2(x) = x^2 + 5$, the derivatives of $3x^2 - 2x + 4$, $(3x^2 - 2x + 4)(x^2 + 5)$, and $\dfrac{3x^2 - 2x + 4}{x^2 + 5}$ can be determined by:

```
>> f1= 3 -2 4];
>> f2=[1 0 5];
```
Creating the vectors coefficients of f_1 and f_2.

```
>> k=polyder(f1)
k =
     6    -2
```
The derivative of f_1 is: $6x - 2$.

```
>> d=polyder(f1,f2)
d =
    12    -6    38   -10
```
The derivative of f_1*f_2 is: $12x^3 - 6x^2 + 38x - 10$.

```
>> [n d]=polyder(f1,f2)
n =
     2    22   -10
d =
     1     0    10     0    25
```
The derivative of $\dfrac{3x^2 - 2x + 4}{x^2 + 5}$ is: $\dfrac{2x^2 + 22x - 10}{x^4 + 10x^2 + 25}$.

8.2 CURVE FITTING

Curve fitting, also called regression analysis, is a process of fitting a function to a set of data points. The function can then be used as a mathematical model of the data. Since there are many types of functions (linear, polynomial, power, exponential, etc.) curve fitting can be a complicated process. Many times there is some idea of the type of function that might fit the given data and the need is only to determine the coefficients of the function. In other situations, where nothing is known about the data, it is possible to make different types of plots which provide information about possible forms of functions that might give a good fit to the data. This section describes some of the basic techniques for curve fitting, and the tools that MATLAB has for this purpose.

8.2.1 Curve Fitting with Polynomials, the `polyfit` Function

Polynomials can be used to fit data points in two ways. In one, the polynomial passes through all the data points, and in the other the polynomial does not necessarily pass through any of the points, but overall gives a good approximation of the data. The two options are described below.

Polynomials that pass through all the points:

When n points (x_i, y_i) are given, it is possible to write a polynomial of degree $n - 1$ that passes through all the points. For example, if two points are given it is possible to write a linear equation in the form of $y = mx + b$ that passes through the points. With three points the equation has the form of $y = ax^2 + bx + c$. With n

points the polynomial has the form: $a_{n-1}x^{n-1} + a_{n-2}x^{n-2} + \ldots + a_1 x + a_0$. The coefficients of the polynomial are determined by substituting each point in the polynomial, and then solving the n equations for the coefficients. As will be shown later in this section, polynomials of high degree might give a large error if they are used to estimate values between data points.

Polynomials that do not necessarily pass through any of the points:

When n points are given, it is possible to write a polynomial of degree less than $n - 1$ that does not necessarily pass through any of the points, but overall approximates the data. The most common method of finding the best fit to data points is the method of least squares. In this method the coefficients of the polynomial are determined by minimizing the sum of the squares of the residuals at all the data points. The residual at each point is defined as the difference between the value of the polynomial and the value of the data. For example, consider the case of finding the equation of a straight line that best fits four data points as shown in Figure 8-1. The points are (x_1, y_1), (x_2, y_2), (x_3, y_3), (x_4, y_4), and the polynomial of the

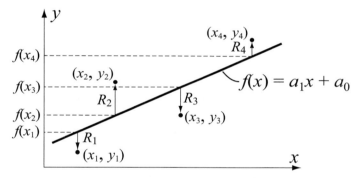

Figure 8-1: Least square fitting of first degree polynomial to four points.

first degree can be written by: $f(x) = a_1 x + a_0$. The residual, R_i at each point, is the difference between the value of the function at x_i and y_i, $R_i = f(x_i) - y_i$. An equation for the sum of the squares of the residuals R_i of all the points is given by:

$$R = [f(x_1) - y_1]^2 + [f(x_2) - y_2]^2 + [f(x_3) - y_3]^2 + [f(x_4) - y_4]^2$$

or, after substituting the equation of the polynomial at each point, by:

$$R = [a_1 x_1 + a_0 - y_1]^2 + [a_1 x_2 + a_0 - y_2]^2 + [a_1 x_3 + a_0 - y_3]^2 + [a_1 x_4 + a_0 - y_4]^2$$

At this stage R is a function of a_1 and a_0. The minimum of R can be determined by taking the partial derivative of R with respect to a_1 and a_0 (2 equations) and equating them to zero.

$$\frac{\partial R}{\partial a_1} = 0 \quad \text{and} \quad \frac{\partial R}{\partial a_0} = 0$$

This results in a system of two equations with two unknowns, a_1 and a_0. The solution of these equations gives the values of the coefficients of the polynomial that best fits the data. The same procedure can be followed with more points and higher order polynomials. More details on the least squares method can be found in books on numerical analysis.

Curve fitting with polynomials is done in MATLAB with the `polyfit` function, which uses the least squares method. The basic form of the `polyfit` function is:

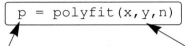

$$p = polyfit(x,y,n)$$

p is the vector of the coefficients of the polynomial that fits the data.

x is a vector with the horizontal coordinate of the data points (independent variable).
y is a vector with the vertical coordinate of the data points (dependent variable).
n is the degree of the polynomial.

For the same set of m points, the `polyfit` function can be used to fit polynomials of any order up to $m - 1$. If $n = 1$ the polynomial is a straight line, if $n = 2$ the polynomial is a parabola, and so on. The polynomial passes through all the points if $n = m - 1$ (the order of the polynomial is one less than the number of points). It should be pointed out here that a polynomial that passes through all the points, or polynomials with higher order, do not necessarily give a better fit overall. High order polynomials can sometimes deviate significantly between the data points.

Figure 8-2 shows how polynomials of different degrees fit the same set of data points. A set of seven points is given by: (0.9, 0.9), (1.5, 1.5), (3, 2.5),

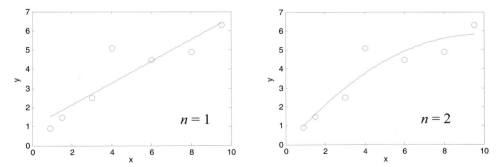

Figure 8-2: Fitting data with polynomials of different order.

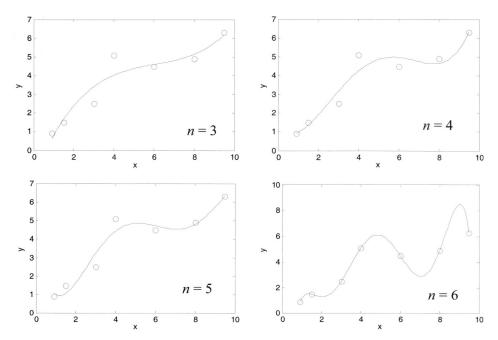

Figure 8-2: Fitting data with polynomials of different order. (Continued)

(4, 5.1), (6, 4.5), (8, 4.9), and (9.5, 6.3). The points are fitted using the `polyfit` function with polynomials of degrees one through six. Each plot in Figure 8-2, shows the same data points, marked with circles, and a curve-fitted line that corresponds to a polynomial with the specified degree. It can be seen that the polynomial with $n = 1$ is a straight line, and with $n = 2$ is a slightly curved line. As the degree of the polynomial increases the line has more bends such that it passes closer to more points. When $n = 6$, which is one less than the number of points, the line passes through all the points. However, between some of the points, the line deviates significantly from the trend of the data.

The script file used to generate one of the plots in Figure 8-2 (the polynomial with $n = 3$) is shown below. Note that in order to plot the polynomial (the line) a new vector xp with small spacing is created. This vector is then used with

```
x=[0.9 1.5 3 4 6 8 9.5];                Create vectors x and y with the
y=[0.9 1.5 2.5 5.1 4.5 4.9 6.3];        coordinates of the data points.
p=polyfit(x,y,3)              Create a vector p using the polyfit function.
xp=0.9:0.1:9.5;         Create a vector xp to be used for plotting the polynomial.
yp=polyval(p,xp)        Create a vector yp with values of the polynomial at each xp.
plot(x,y,'o',xp,yp)             A plot of the 7 points and the polynomial.
xlabel('x'); ylabel('y')
```

the function `polyval` to create a vector `yp` with the value of the polynomial for each element of `xp`.

When the script file is executed, the following vector `p` is displayed in the Command Window.

```
p =
      0.0220    -0.4005     2.6138    -1.4158
```

This means that the polynomial of the third degree in Figure 8-2 has the form:
$0.022x^3 - 0.4005x^2 + 2.6138x - 1.4148$.

8.2.2 Curve Fitting with Functions Other than Polynomials

Many situations in science and engineering require fitting functions that are not polynomials to given data. Theoretically, any function can be used to model data within some range. For a particular data set, however, some functions provide a better fit than others. In addition, determining the best fit coefficients of some functions can be more difficult than others. This section covers curve fitting with power, exponential, logarithmic, and reciprocal functions, which are commonly used. The forms of these functions are:

$$y = bx^m \qquad \text{(power function)}$$
$$y = be^{mx} \ \text{ or } \ y = b10^{mx} \qquad \text{(exponential function)}$$
$$y = m\ln(x) + b \ \text{ or } \ y = m\log(x) + b \quad \text{(logarithmic function)}$$
$$y = \frac{1}{mx + b} \qquad \text{(reciprocal function)}$$

All of these functions can easily be fitted to given data with the `polyfit` function. This is done by rewriting the functions in a form that can be fitted with a linear polynomial ($n = 1$) which has the form:

$$y = mx + b$$

The logarithmic function is already in this form, and the power, exponential and reciprocal equations can be rewritten as:

$$\ln(y) = m\ln(x) + \ln b \qquad \text{(power function)}$$
$$\ln(y) = mx + \ln(b) \ \text{ or } \ \log(y) = mx + \log(b) \quad \text{(exponential function)}$$
$$\frac{1}{y} = mx + b \qquad \text{(reciprocal function)}$$

These equations describe a linear relationship between $\ln(y)$ and $\ln(x)$ for the power function, between $\ln(y)$ and x for the exponential function, between y and $\ln(x)$ or $\log(x)$ for the logarithmic function, and between $1/y$ and x for the reciprocal function. This means that the `polyfit(x,y,1)` function can be used to determine the best fit constants m and b for best fit if, instead of x and y, the following arguments are used.

Function		`polyfit` **function form**
power	$y = bx^m$	p=polyfit(log(x),log(y),1)
exponential	$y = be^{mx}$ or	p=polyfit(x,log(y),1) or
	$y = b10^{mx}$	p=polyfit(x,log10(y),1)
logarithmic	$y = m\ln(x) + b$ or	p=polyfit(log(x),y,1) or
	$y = m\log(x) + b$	p=polyfit(log10(x),y,1)
reciprocal	$y = \dfrac{1}{mx + b}$	p=polyfit(x,1./y,1)

The result of the `polyfit` function is assigned to p, which is a two element vector. The first element, p(1), is the constant m, and the second element, p(2), is b for the logarithmic and reciprocal functions, $\ln(b)$ or $\log(b)$ for the exponential function, and $\ln(b)$ for the power function ($b = e^{p(2)}$ or $b = 10^{p(2)}$ for the exponential function and $b = e^{p(2)}$ for the power function).

For given data it is possible to foresee, to some extent, which of the functions has the potential for providing a good fit. This is done by plotting the data using different combinations of linear and logarithmic axes. If the data points in one of the plots appear to fit a straight line, the corresponding function can provide a good fit according to the list below.

x axis	**y axis**	**Function**
linear	linear	linear $y = mx + b$
logarithmic	logarithmic	power $y = bx^m$
linear	logarithmic	exponential $y = be^{mx}$ or $y = b10^{mx}$
logarithmic	linear	logarithmic $y = m\ln(x) + b$ or $y = m\log(x) + b$
linear	linear (plot $1/y$)	reciprocal $y = \dfrac{1}{mx + b}$

Other considerations when choosing a function are:

- Exponential functions can not pass through the origin.
- Exponential functions can only fit data with all positive y's, or all negative y's.
- Logarithmic functions can not model $x = 0$, or negative values of x.
- For the power function $y = 0$ when $x = 0$.
- The reciprocal equation can not model $y = 0$.

The following example illustrates the process of fitting a function to a set of data points.

Sample Problem 8-2: Fitting an equation to data points

The following data points are given. Determine a function $w = f(t)$ (t is the independent variable, w is the dependent variable) with a form discussed in this section that best fits the data.

t	0.0	0.5	1.0	1.5	2.0	2.5	3.0	3.5	4.0	4.5	5.0
w	6.00	4.83	3.70	3.15	2.41	1.83	1.49	1.21	0.96	0.73	0.64

Solution

The data is first plotted with linear scales on both axes. The figure, shown on the right, indicates that a linear function will not give the best fit since the points do not appear to line up along a straight line. From the other possible functions, the logarithmic function is excluded since in the first point $t = 0$, and the power function is excluded since at $t = 0$, $w \neq 0$. To check if
the other two functions (exponential and reciprocal) might give a better fit, two additional plots, shown below, are made. The plot on the left has a log scale on the vertical axis, and linear horizontal axis. In the plot on the right both axes have linear scales, and the quantity $1/w$ is plotted on the vertical axis.

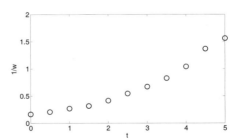

In the left figure the data points appear to line up along a straight line. This indicates that an exponential function in the form $y = be^{mx}$ can give a good fit to the data. A program in a script file that determines the constants b and m, and plots the data points and the function is given below.

```
t=0:0.5:5;      Create vectors t and w with the coordinates of the data points.
w=[6 4.83 3.7 3.15 2.41 1.83 1.49 1.21 0.96 0.73 0.64];
p=polyfit(t,log(w),1);  Use the polyfit function with t and log(w).
```

```
m=p(1)
b=exp(p(2))            Determine the coefficient b.
tm=0:0.1:5;        Create a vector tm to be used for plotting the polynomial.
wm=b*exp(m*tm);       Calculate the function value at each element of tm.
plot(t,w,'o',tm,wm)        Plot the data points and the function.
```

When the program is executed, the values of the constants *m* and *b* are displayed in the Command Window.

```
m =
    -0.4580
b =
    5.9889
```

The plot, generated by the program, that shows the data points and the function is (axes labels were added with the Plot Editor):

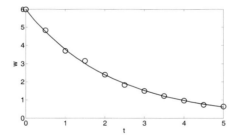

It should be pointed out here that in addition to the power, exponential, logarithmic, and reciprocal functions that are discussed in this section, many other functions can be written in a form suitable for curve fitting with the `polyfit` function. One example where a function of the form $y = e^{(a_2x^2 + a_1x + a_0)}$ is fitted to data points using the `polyfit` function with a third order polynomial is described in Sample Problem 8-7.

8.3 INTERPOLATION

Interpolation is estimation of values between data points. MATLAB has interpolation functions that are based on polynomials, which are described in this section, and on Fourier transformation, which are outside the scope of this book. In one-dimensional interpolation each point has one independent variable (*x*) and one dependent variable (*y*). In two-dimensional interpolation each point has two independent variables (*x* and *y*) and one dependent variable (*z*).

One-dimensional interpolation:

If only two data points exist, the points can be connected with a straight line and a linear equation (polynomial of first order) can be used to estimate values between the points. As was discussed in the previous section, if three (or four) data points exist, a second (or a third) order polynomial can be determined to pass through the points and then be used to estimate values between the points. As the number of points increases, a higher order polynomial is required for the polynomial to pass through all the points. Such a polynomial, however, will not necessarily give a good approximation of the values between the points. This is illustrated in Figure 8-2 with $n = 6$.

A more accurate interpolation can be obtained if instead of considering all the points in the data set (by using one polynomial that passes through all the points), only a few data points in the neighborhood where the interpolation is needed are considered. In this method, called spline interpolation, many low order polynomials are used, where each is valid only in a small domain of the data set.

The simplest method of spline interpola-
tion is called linear spline interpolation. In this method, shown on the right, every two adjacent points are connected with a straight line (a polynomial of first degree). The equation of a straight line that passes through two adjacent points (x_i, y_j) and (x_{i+1}, y_{j+1}) and can be used to calculate the value of y for any x between the points is given by:

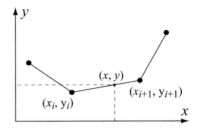

$$y = \frac{y_{i+1} - y_i}{x_{i+1} - x_i}x + \frac{y_i x_{i+1} - y_{i+1}x_i}{x_{i+1} - x_i}$$

In a linear interpolation the line between two data points has a constant slope, and there is a change in the slope at every point. A smoother interpolation curve can be obtained by using quadratic, or cubic polynomials. In these methods, called quadratic splines or cubic splines, a second, or third order polynomial is used to interpolate between every two points. The coefficients of the polynomial are determined by using data from additional points that are adjacent to the two data points. The theoretical background for the determination of the constants of the polynomials is beyond the scope of this book and can be found in books on numerical analysis.

One-dimensional interpolation in MATLAB is done with the `interp1` (last character is the number one) function, which has the form:

$$\boxed{\text{yi = interp1(x,y,xi,'method')}}$$

yi is the interpolated value.

x is a vector with the horizontal coordinate of the input data points (independent variable).
y is a vector with the vertical coordinate of the input data points (dependent variable).
xi is the horizontal coordinate of the interpolation point (independent variable).

Method of interpolation, typed as a string (optional).

- The vector x must be monotonic (the elements in ascending or descending order).

- xi can be a scalar (interpolation of one point) or a vector (interpolation of many points). Respectively, yi is a scalar or a vector with the corresponding interpolated values.

- MATLAB can do the interpolation using one of several methods that can be specified. These methods include:

  'nearest' returns the value of the data point that is nearest to the interpolated point.
  'linear'   uses linear spline interpolation.
  'spline'   uses cubic spline interpolation.
  'pchip'   uses piecewise cubic Hermite interpolation, also called 'cubic'

- When the 'nearest' and the 'linear' methods are used, the value(s) of xi must be within the domain of x. If the 'spline' or the 'pchip' methods are used, xi can have values outside the domain of x and the function interp1 performs extrapolation.

- The 'spline' method can give large errors if the input data points are non-uniform such that some points are much closer together than others.

- Specification of the method is optional. If no method is specified the default is 'linear'.

Sample Problem 8-3: Interpolation

The following data points, which are points of the function $f(x) = 1.5^x \cos(2x)$, are given. Use linear, spline, and pchip interpolation methods to calculate the value of y between the points. Make a figure for each of the interpolation methods. In the figure show the points, a plot of the function, and a curve that corresponds to the interpolation method.

x	0	1	2	3	4	5
y	1.0	−0.6242	−1.4707	3.2406	−0.7366	−6.3717

Solution

The following is a program written in a script file that solves the problem:

```
x=0:1.0:5;              Create vectors x and y with coordinates of the data points.
y=[1.0 -0.6242 -1.4707 3.2406 -0.7366 -6.3717];
xi=0:0.1:5;                  Create vector xi with points for interpolation.
yilin=interp1(x,y,xi,'linear');  Calculate y points from linear interpolation.
yispl=interp1(x,y,xi,'spline');  Calculate y points from spline interpolation.
yipch=interp1(x,y,xi,'pchip');   Calculate y points from pchip interpolation.
yfun=1.5.^xi.*cos(2*xi);         Calculate y points from the function.
subplot(1,3,1)
plot(x,y,'o',xi,yfun,xi,yilin,'--');
subplot(1,3,2)
plot(x,y,'o',xi,yfun,xi,yispl,'--');
subplot(1,3,3)
plot(x,y,'o',xi,yfun,xi,yipch,'--');
```

The three figures generated by the program are shown below (axes labels were added with the Plot Editor). The data points are marked with circles, the interpolation curves are plotted with dashed lines, and the function is shown with a solid line. The left figure shows the linear interpolation, the middle is the spline, and the figure on the right shows the pchip interpolation.

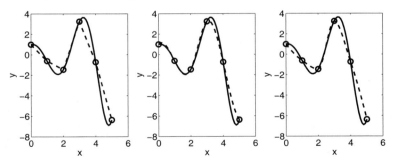

8.4 THE BASIC FITTING INTERFACE

The basic fitting interface is a tool that can be used to perform curve fitting and interpolation interactively. By using the interface the user can:

- Curve fit the data points with polynomials of various degrees up to 10, and with spline and Hermite interpolation methods.

- Plot the various fits on the plot simultaneously so that they can be compared.

- Plot the residuals of the various polynomial fits and compare the norm of the residuals.

- Calculate the value of specific points with the various fits.

- Add the equations of the polynomials to the plot.

To activate the basic fitting interface, the user first has to make a plot of the data points. Then, the interface is activated by selecting **Basic Fitting** in the **Tools** menu, as shown on the right. This opens the Basic Fitting Window, shown in Figure 8-3. When the window first opens, only one panel (the **Plot fits** panel) is visible. The window can be extended to show a second panel (the **Numerical results** panel) by clicking on the → button. One click adds the first section of the panel, and a second click makes the window look as it is shown in Figure 8-3. The window can be reduced back by clicking on the ← button. The first two items in the Basic Fitting Window are related to the selection of the data points:

Select data: Used to select a specific set of data points for curve fitting in a figure that has more than one set of data points. Only one set of data points can be curve-fitted at a time, but multiple fits can be performed simultaneously on the same set.

Center and scale x data: When this box is checked, the data is centered at zero mean and scaled to unit standard deviation. This might be needed in order to improve the accuracy of numerical computation.

The next four items are in the **Plot fits** panel and are related to the display of the fit:

Check to display fits on figure: The user selects the fits to be displayed in the figure. The selection includes interpolation with spline interpolant (interpolation method) that uses the spline function, interpolation with Hermite interpolant that uses the pchip function, and polynomials of various degrees that use the

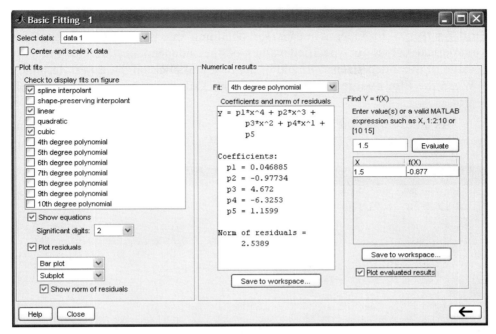

Figure 8-3: The Basic Fitting Window.

`polyfit` function. Several fits can be selected and displayed simultaneously.

Show equations: When this box is checked, the equations of the polynomials that were selected for the fit are displayed in the figure. The equations are displayed with the number of significant digits that is selected in the adjacent significant digits menu.

Plot residuals: When this box is checked, a plot that shows the residual at each data point is created (residuals are defined in Section 8.2.1). By selecting the proper choices in the adjacent windows, the plot can be a bar plot, a scatter plot, or a line plot, and can be displayed as a subplot in the same Figure Window that has the plot of the data points, or as a separate plot in a different Figure Window.

Show norm of residuals: When this box is checked, the norm of the residuals is displayed in the plot of the residuals. The norm of the residual is a measure of the quality of the fit. A smaller norm corresponds to a better fit.

The next three items are in the **Numerical results** panel. They provide the numerical information of one fit, independently of the fits that are displayed:

Fit: The user selects the fit he wants to examine numerically. The fit is shown on the plot only if it is selected in the **Plot fit** panel.

Coefficients and norm of residuals: Displays the numerical results for the polynomial fit that is selected in the **Fit** menu. It includes the coefficients of the polynomial and the norm of the residuals. The results can be saved by clicking on the **Save to workspace** button.

Find Y = f(x):   Provides a means for obtaining interpolated (or extrapolated) numerical values for specified values of the independent variable. Enter the value of the independent variable in the box, and click on the **Evaluate** button. When the **Plot evaluated result** box is checked, the point is displayed on the plot.

    As an example, the basic fitting interface is used for fitting the data points from Sample Problem 8-3. The Basic Fitting Window is the one shown in Figure

Figure 8-4: A Figure Window modified by the Basic Fitting Interface.

8-3, and the corresponding Figure Window is shown in Figure 8-4. The Figure Window includes a plot of the points, one interpolation fit (spline), two polynomial fits (linear and cubic), a display of the equations of the polynomial fits, and a mark of the point $x = 1.5$ that is entered in the **Find Y = f(x)** box of the Basic Fitting Window. The Figure Window also includes a plot of the residuals of the polynomial fits and a display of their norm.

8.5 *EXAMPLES OF MATLAB APPLICATIONS*

Sample Problem 8-4:   Determining wall thickness of a box

The outside dimensions of a rectangular
box (bottom and four sides, no top), made
of aluminum, are 24 by 12 by 4 inches. The
wall thickness of the bottom and the sides
is x. Derive an expression that relates the
weight of the box and the wall thickness x.
Determine the thickness x for a box that
weighs 15 lb. The specific weight of aluminum is 0.101 lb/in^3.

Solution

The volume of the aluminum V_{Al} is calculated from the weight W of the box by:

$$V_{Al} = \frac{W}{\gamma}$$

where γ is the specific weight. The volume of the aluminum from the dimensions
of the box is given by:

$$V_{Al} = 24 \cdot 12 \cdot 4 - (24 - 2x)(12 - 2x)(4 - x)$$

where the inside volume of the box is subtracted from the outside volume. This
equation can be rewritten as:

$$(24 - 2x)(12 - 2x)(4 - x) + V_{Al} - (24 \cdot 12 \cdot 4) = 0$$

which is a third degree polynomial. A root of this polynomial is the required
thickness x. A program in a script file that determines the polynomial and solves
for the roots is:

```
W=15; gamma=0.101;                    Assign W and gamma.
VAlum=W/gamma;                        Calculate the volume of the aluminum.
a=[-2  24];                           Assign the polynomial 24 – 2x to a.
b=[-2  12];                           Assign the polynomial 12 – 2x to b.
c=[-1  4];                            Assign the polynomial 4 – x to c.
Vin=conv(c, conv(a,b));              Multiply the three polynomials above.
polyeq=[0 0 0 (VAlum-24*12*4)]+Vin   Add V_Al – 24*12*4 to Vin.
x=roots(polyeq)                       Determine the roots of the polynomial.
```

Note in the second to last line, that in order to add the quantity $V_{Al} - (24 \cdot 12 \cdot 4)$ to
the polynomial Vin it has to be written as a polynomial of the same order as Vin
(Vin is a polynomial of third order). When the program (saved as
Chap8SamPro4) is executed, the coefficients of the polynomial and the value of x
are displayed:

```
>> Chap8SamPro4
polyeq =
  -4.0000   88.0000  -576.0000   148.5149
x =
  10.8656 + 4.4831i
  10.8656 - 4.4831i
   0.2687
```

The polynomial is:

$-4x^3 + 88x^2 - 576x + 148.515$.

The polynomial has one real root, $x = 0.2687$ in. which is the thickness of the aluminum wall.

Sample Problem 8-5:   Floating height of a buoy

An aluminum thin-walled sphere is used as a marker buoy. The sphere has a radius of 60 cm, and a wall thickness of 12 mm. The density of aluminum is ρ_{Al} = 2690 kg/m³. The buoy is placed in the ocean where the density of the water is 1030 kg/m³. Determine the height h between the top of the buoy and the surface of the water.

Solution

According to Archimedes law, the buoyancy force applied to an object that is placed in a fluid is equal to the weight of the fluid that is displaced by the object. Accordingly, the aluminum sphere will be at such a depth that the weight of the sphere is equal to the weight of the fluid displaced by the part of the sphere that is submerged in the water.

The weight of the sphere is given by:

$$W_{sph} = \rho_{Al} V_{Al} g = \rho_{Al} \frac{4}{3} \pi (r_o^3 - r_i^3) g$$

where V_{Al} is the volume of the aluminum, r_o and r_i are the outside and inside radii of the sphere, respectively, and g is the gravitational acceleration.

The weight of the water that is displaced by the spherical cap that is submerged in the water is given by:

$$W_{wtr} = \rho_{wtr} V_{wtr} g = \rho_{wtr} \frac{1}{3} \pi (2r_o - h)^2 (r_o + h) g$$

Setting the two weights equal to each other gives the following equation:

$$h^3 - 3r_o h^2 + 4r_o^3 - 4 \frac{\rho_{Al}}{\rho_{wtr}} (r_o^3 - r_i^3) = 0$$

The last equation is a third degree polynomial for h. The root of the polynomial is the answer.

A solution with MATLAB is obtained by writing the polynomials and using the `roots` function to determine the value of h. This is done in the following script file:

```
rout=0.60; rin=0.588;           Assign the radii to variables.
rhoalum=2690; rhowtr=1030;      Assign the densities to variables.
a0=4*rout^3-4*rhoalum*(rout^3-rin^3)/rhowtr;  Assign the coefficient a0.
p = [1 -3*rout 0 a0];     Assign the coefficient vector of the polynomial.
h = roots(p)                Calculate the roots of the polynomial.
```

When the script file is executed in the Command Window, as shown below, the answer is three roots, since the polynomial was of the third degree. The only answer that is physically possible is the second, where $h = 0.9029$ m.

```
>> Chap8SamPro5

h =
     1.4542        The polynomial has three roots. The only one that is
     0.9029        physically possible for the problem is 0.9029 m.
    -0.5570
```

Sample Problem 8-6:   Determining the size of a capacitor

An electrical capacitor has an unknown capacitance. In order to determine its capacitance it is connected to the circuit shown. The switch is first connected to B and the capacitor is charged. Then, the switch is switched to A and the capacitor discharges through the resistor. As the capacitor is dis-

charging, the voltage across the capacitor is measured for 10 s in intervals of 1 s. The recorded measurements are given in the table below. Plot the voltage as a function of time and determine the capacitance of the capacitor by fitting an exponential curve to the data points.

t (s)	1	2	3	4	5	6	7	8	9	10
V (V)	9.4	7.31	5.15	3.55	2.81	2.04	1.26	0.97	0.74	0.58

Solution

When a capacitor discharges through a resistor, the voltage of the capacitor as a function of time is given by:

$$V = V_0 e^{(-t)/(RC)}$$

Where V_0 is the initial voltage, R the resistance of the resistor, and C the capacitance of the capacitor. As was explained in Section 8.2.2 the exponential function can be written as a linear equation for $\ln(V)$ and t in the form:

$$\ln(V) = \frac{-1}{RC}t + \ln(V_0)$$

This equation which has the form $y = mx + b$ can be fitted to the data points by using the `polyfit(x,y,1)` function with t as the independent variable x and $\ln(V)$ as the dependent variable y. The coefficients m and b determined by the `polyfit` function are then used to determine C and V_0 by:

$$C = \frac{-t}{Rm} \quad \text{and} \quad V_0 = e^b$$

The following program written in a script file determines the best fit exponential function to the data points, determines C and V_0, and plots the points and the fitted function.

```
R=2000;                                                    Define R.
t=1:10;                               Assign the data points to vectors t and v.
v=[9.4 7.31 5.15 3.55 2.81 2.04 1.26 0.97 0.74 0.58];
p=polyfit(t,log(v),1);      Use the polyfit function with t and log(v).
C=-1/(R*p(1))             Calculate C from p(1), which is m in the equation.
V0=exp(p(2))             Calculate V0 from p(2), which is b in the equation.
tplot=0:0.1:10;          Create vector tplot of time for plotting the function.
vplot=V0*exp(-tplot./(R*C));    Create vector vplot for plotting the function.
plot(t,v,'o',tplot,vplot)
```

When the script file is executed (saved as Chap8SamPro6) the values of C and V_0 are displayed in the Command Window as shown below:

```
>> Chap8SamPro6
C =                              The capacitance of the capacitor is 1600 μF.
    0.0016
V0 =
    13.2796
```

The program creates also the following plot (axes labels were added to the plot using the Plot Editor):

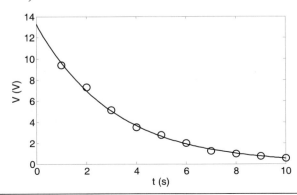

Sample Problem 8-7: Temperature dependence of viscosity

Viscosity, μ, is a property of gases and fluids that characterizes their resistance to flow. For most materials viscosity is highly sensitive to temperature. Below is a table that gives the viscosity of SAE 10W oil at different temperatures (Data from B.R. Munson, D.F. Young, and T.H. Okiishi, "Fundamental of Fluid Mechanics," 4th Edition, John Wiley and Sons, 2002). Determine an equation that can be fitted to the data.

T (°C)	−20	0	20	40	60	80	100	120
μ (N s/m²) ($\times 10^{-5}$)	4	0.38	0.095	0.032	0.015	0.0078	0.0045	0.0032

Solution

To examine what type of equation might provide a good fit to the data, μ is plotted as a function of T (absolute temperature) with linear scale for T and logarithmic scale for μ. The plot, shown on the right indicates that the data points do not appear to line up along a straight line. This means that a simple exponential function of the form $y = be^{mx}$, which models a straight line in these axes, will not

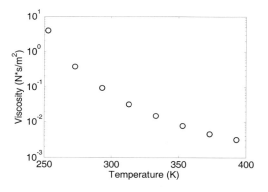

provide the best fit. Since the points in the figure appear to lie along a curved line, a function that can possibly have a good fit to the data is:

$$\ln(\mu) = a_2 T^2 + a_1 T + a_0$$

This function can be fitted to the data by using MATLAB `polyfit(x,y,2)` function (second degree polynomial), where the independent variable is T and the dependent variable is $\ln(\mu)$. The equation above can be solved for μ to give the viscosity as a function of temperature:

$$\mu = e^{(a_2 T^2 + a_1 T + a_0)} = e^{a_0} e^{a_1 T} e^{a_2 T^2}$$

The following program determines the best fit of the function, and creates a plot that displays the data points and the function.

```
T=[-20:20:120];
mu=[4 0.38 0.095 0.032 0.015 0.0078 0.0045 0.0032];
TK=T+273;
p=polyfit(TK,log(mu),2)
Tplot=273+[-20:120];
```

```
muplot = exp(p(1)*Tplot.^2 + p(2)*Tplot + p(3));
semilogy(TK,mu,'o',Tplot,muplot)
```

When the program executes (saved as Chap8SamPro7), the coefficients that are determined by the `polyfit` function are displayed in the Command Window (shown below) as three elements of the vector p.

```
>> Chap8SamPro7
p =
    0.0003   -0.2685    47.1673
```

With these coefficients the viscosity of the oil as a function of temperature is:

$$\mu = e^{(0.0003\,T^2 - 0.2685\,T + 47.1673)} = e^{47.1673}\,e^{(-0.2685)\,T}\,e^{0.0003\,T^2}$$

The plot that is generated shows that the equation correlates well to the data points (axes labels were added with the Plot Editor).

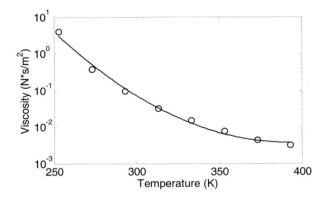

8.6 PROBLEMS

1. Plot the polynomial $y = 0.02x^4 - 0.75x^3 + 12.5x - 2$ in the domain $-6 \le x \le 6$. First create a vector for x, next use the `polyval` function to calculate y, and then use the `plot` function.

2. Divide the polynomial $12x^6 + 21x^5 - 11x^4 - 14x^3 + 18x^2 + 28x - 4$ by the polynomial $4x^2 + 7x - 1$.

3. Divide the polynomial $4x^4 + 6x^3 - 2x^2 - 5x + 3$ by the polynomial $x^2 + 4x + 2$.

4. A cylindrical stainless-steel fuel tank has an outside diameter of 40 cm and a length of 70 cm. The side of the cylinder and the bottom and top ends have a thickness of x. Determine x if the mass of the tank is 18 kg. The density of stainless steel is 7920 kg/m³.

5. A rectangular piece of cardboard, 40 inches long by 22 inches wide, is used for making a rectangular box (open top) by cutting out squares of x by x from the corners and folding up the sides.

   a) Create a polynomial expression for the volume V in terms of x.
   b) Make a plot of V versus x.
   c) Determine x if the volume of the box is 1000 in³.
   d) Determine the value of x that corresponds to the box with the largest possible volume, and determine that volume.

6. A gas tank has the geometry shown in the figure (two frustum end caps that are attached to a circular cylinder). The radius of the cylinder is R and its length is $2R$. Each frustum has a base with radius R, and a top with radius 0.25 m, and a height of 0.4m. Determine R if the volume of the tank is 1.7 m³.

7. Write a user-defined function that adds or subtracts two polynomials of any order. Name the function p=polyadd(p1,p2,operation). The first two input arguments p1 and p2 are the vectors of the coefficients of the two polynomials. (If the two polynomials are not of the same order the function adds the necessary zero elements to the shorter vector.) The third input argument operation is a string that can be either 'add' or 'sub', for adding or subtracting the polynomials, respectively, and the output argument is the resulting polynomial.

   Use the function to add and subtract the following polynomials:
   $f_1(x) = x^5 - 7x^4 + 11x^3 - 4x^2 - 5x - 2$ and $f_2(x) = 9x^2 - 10x + 6$.

8. Write a user-defined function that calculates the maximum (or minimum) of a quadratic equation of the form:
$$f(x) = ax^2 + bx + c$$

Name the function $[x,y,w] = \text{maxormin}(a,b,c)$. The input arguments are the coefficients a, b, and c. The output arguments are x the coordinate of the maximum (or minimum), y the maximum (or minimum) value, and w which is equal to 1 if y is a maximum, and equal to 2 if y is a minimum.

Use the function to determine the maximum or minimum of the following functions:

a) $f(x) = 3x^2 - 7x + 14$ b) $f(x) = -5x^2 - 11x + 15$

9. A cylinder of radius r and height h is constructed inside a cone with base radius $R = 10$ in. and height of $H = 30$ in., as shown in the figure.

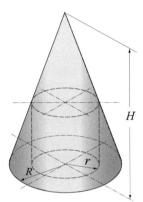

a) Create a polynomial expression for the volume V of the cylinder in terms of r.
b) Make a plot of V versus r.
c) Determine r if the volume of the cylinder is 800 in^3.
d) Determine the value of r that corresponds to the cylinder with the largest possible volume, and determine that volume.

10. The van der Waals equation gives a relationship between the pressure p (in atm.), volume V (in L), and temperature T (in K) for a real gas:

$$p = \frac{nRT}{V - nb} - \frac{n^2 a}{V^2}$$

where n is the number of moles, $R = 0.08206$ (L atm)/(mole K) is the gas constant, and a (in L^2 atm/mole^2), and b (in L/mole) are material constants. The equation can be easily used for calculating p (given T and V), or T (given p and V). The equation is not as readily solved for V when p and T are given, since it is nonlinear in V. One useful way to solve for V is by rewriting the equation as a third-order polynomial:

$$V^3 - \left(nb + \frac{nRT}{p}\right)V^2 + \frac{n^2 a}{p}V - \frac{n^3 ab}{p} = 0$$

and calculating the root of the polynomial.

Write a user-defined function that calculates V for given p, T, n, a, and b. For function name and arguments use $V=\text{waals}(p,T,n,a,b)$. The function calculates V by using MATLAB's built-in function roots. Note that the solution of the polynomial can have non-real (complex) roots. The output argument V in waals should be the physically realistic solution (positive and real). (MATLAB's built-in function $\text{imag}(x)$ can be used for determining which root is real.)

Use the user-defined function to calculate V for $p = 30$ atm, $T = 300$ K, $n = 1.5$, $a = 1.345$ L^{2}atm/mole^2, $b = 0.0322$ L/mole.

11. The following points are given:

x	−6	−3.5	−2.5	−1	0	1.5	2.2	4	5.2	6	8
y	0.3	0.4	1.1	3.6	3.9	4.5	4.2	3.5	4.0	5.3	6.1

a) Fit the data with a first-order polynomial. Make a plot of the points and the polynomial.

b) Fit the data with a second-order polynomial. Make a plot of the points and the polynomial.

c) Fit the data with a fourth-order polynomial. Make a plot of the points and the polynomial.

d) Fit the data with a tenth-order polynomial. Make a plot of the points and the polynomial.

12. The population of India from the year 1940 to the year 2000 is given in the following table:

Year	1910	1930	1940	1950	1960	1970	1980	1990	2000
Population (millions)	249	277	316	350	431	539	689	833	1014

a) Determine the exponential function that best fits the data. Use the function to estimate the population in 1975.

b) Curve fit the data with a quadratic equation (second-order polynomial). Use the function to estimate the population in 1975.

c) Fit the data with linear and spline interpolations. Estimate the population in 1975 with linear and spline interpolations.

In each part make a plot of the data points (circle markers) and the curve fitting or the interpolation curves. Note that part *c* has two interpolation curves. The actual population in India in 1955 was 600.8 million.

13. The standard air density, D (average of measurements made) at different heights, h, from sea level up to a height of 33 km is given below.

h (km)	0	3	6	9	12	15
D (kg/m^3)	1.2	0.91	0.66	0.47	0.31	0.19
h (km)	18	21	24	27	30	33
D (kg/m^3)	0.12	0.075	0.046	0.029	0.018	0.011

a) Make the following four plots of the data points (density as a function of height): (1) both axes with linear scale, (2) h with log axis, D with linear axis, (3) h with linear axis D with log axis, (4) both with log axes. According to the plots choose a function (linear, power, exponential, or logarithmic) that can best fit the data points and determine the coefficients of the function.

b) Plot the function and the points using linear axes.

14. Write a user-defined function that fits data points to an exponential function of the form: $y = be^{mx}$. Name the function $[b,m] = \text{expofit}(x,y)$, where the input arguments x and y are vectors with the coordinates of the data points, and the output arguments b and m are the constants of the fitted exponential equation. Use expofit to fit the data below. Make a plot that shows the data points and the function.

x	0.6	2.1	3.1	5.1	6.2	7.6
y	0.9	9.1	24.7	58.2	105	222

15. The aerodynamic drag force F_D that is applied to a car is given by:

$$F_D = \frac{1}{2}\rho C_D A v^2$$

where $\rho = 1.2$ kg/m^3 is the air density, C_D is the drag coefficient, A is the projected front area of the car, and v is the speed of the car (in units of m/s) relative to the wind. The product $C_D A$ characterizes the air resistance of a car. (At speeds above 70 km/h the aerodynamic drag force is typically more than half of the total resistance to motion.) Data obtained in a wind tunnel test is displayed in the table. Use the data to determine the product $C_D A$ for the tested car by using curve fitting. Make a plot of the data points and the curve fitted equation.

v (km/h)	20	40	60	80	100	120	140	160
F_D (N)	10	50	109	180	300	420	565	771

16. The relationship between two variables P and t is known to be:

$$P = \frac{5}{m\sqrt{t} + b}$$

The following data points are known:

t	1	2	3	4	5	6	7
P	8.9	3.4	2.1	2.0	1.6	1.3	1.2

Determine the constants m and b by curve fitting the equation to the data points. Make a plot of P versus t. In the plot show the data points with markers and the curve-fitted equation with a solid line. (The curve fitting can be done by writing the reciprocal of the equation, and using a first-order polynomial.)

17. The yield strength, σ_y, of many metals depends on the size of the grains. For these metals the relationship between the yield stress and the average grain diameter d can be modeled by the Hall-Petch equation:

$$\sigma_y = \sigma_0 + kd^{\left(\frac{-1}{2}\right)}$$

The following are results from measurements of average grain diameter and yield stress.

d (mm)	0.005	0.009	0.016	0.025	0.040	0.062	0.085	0.110
σ_y (MPa)	205	150	135	97	89	80	70	67

a) Using curve fitting, determine the constants σ_0 and k in the Hall-Petch equation for this material. Using the constants determine with the equation the yield stress of material with grain size of 0.05 mm. Make a plot that shows the data points with circle markers and the Hall-Petch equation with a solid line.

b) Use linear interpolation to determine the yield stress of material with grain size of 0.05 mm. Make a plot that shows the data points with circle markers and linear interpolation with a solid line.

c) Use cubic interpolation to determine the yield stress of material with grain size of 0.05 mm. Make a plot that shows the data points with circle markers and cubic interpolation with a solid line.

18. The ideal gas equation relates the volume, pressure, temperature, and the quantity of a gas by:

$$V = \frac{nRT}{P}$$

where V is the volume in liters, P is the pressure in atm, T is the temperature in degrees K, n is the number of moles, and R is the gas constant.

An experiment is conducted for determining the value of the gas constant R. In the experiment 0.05 mol of gas is compressed to different volumes by applying pressure to the gas. At each volume the pressure and temperature of the gas is recorded. Using the data given below, determine R by plotting V versus T/P, and fitting the data points with a linear equation.

V (L)	0.75	0.65	0.55	0.45	0.35
T (°C)	25	37	45	56	65
P (atm)	1.63	1.96	2.37	3.00	3.96

19. Viscosity is a property of gases and fluids that characterizes their resistance to flow. For most materials viscosity is highly sensitive to temperature. For gases, the variation of viscosity with temperature is frequently modeled by an equation which has the form:

$$\mu = \frac{CT^{3/2}}{T + S}$$

where μ is the viscosity, T the absolute temperature, and C and S are empirical constants. Below is a table that gives the viscosity of air at different temperatures (data from B.R. Munson, D.F. Young, and T.H. Okiishi, "Fundamental of Fluid Mechanics," 4th Edition, John Wiley and Sons, 2002).

T (°C)	−20	0	40	100	200	300	400	500	1000
μ (N s/m²) $(\times 10^{-5})$	1.63	1.71	1.87	2.17	2.53	2.98	3.32	3.64	5.04

Determine the constants C and S by curve fitting the equation to the data points. Make a plot of viscosity versus temperature (in °C). In the plot show the data points with markers and the curve-fitted equation with a solid line.

The curve fitting can be done by rewriting the equation in the form:

$$\frac{T^{3/2}}{\mu} = \frac{1}{C}T + \frac{S}{C}$$

and using a first-order polynomial.

Chapter 9
Three-Dimensional Plots

Three-dimensional (3-D) plots can be a useful way to present data that consists of more than two variables. MATLAB provides various options for displaying three-dimensional data. They include line and wire, surface, mesh plots, and many others. The plots can also be formatted to have a specific appearance and special effects. Many of the three-dimensional plotting features are described in this chapter. Additional information can be found in the Help Window under **Plotting and Data Visualization**.

In many ways this chapter is a continuation of Chapter 5 where two-dimensional plots were introduced. The 3-D plots are presented in a separate chapter since not all MATLAB users use them. In addition, it was felt that for new users of MATLAB it is better to practice 2-D plotting first and learn the material in Chapters 6–8 before attempting 3-D plotting. It is assumed in the rest of this chapter that the reader is familiar with 2-D plotting.

9.1 LINE PLOTS

A three-dimensional line plot is a line that is obtained by connecting points in a three-dimensional space. A basic 3-D plot is created with the `plot3` command, which is very similar to the `plot` command, and has the form:

```
plot3(x,y,z,'line specifiers','PropertyName',property value)
```

x, y, and z are vectors of the coordinates of the points.

(Optional) Specifiers that define the type and color of the line and markers.

(Optional) Properties with values that can be used to specify the line width, and marker's size and edge and fill colors.

- The three vectors with the coordinates of the data points must have the same number of elements.

- The line specifiers, properties, and property values are the same as in 2-D plots (see Section 5.1).

For example, if the coordinates x, y, and z are given as a function of the parameter t by:

$$x = \sqrt{t}\sin(2t)$$
$$y = \sqrt{t}\cos(2t)$$
$$z = 0.5t$$

A plot of the points for $0 \le t \le 6\pi$ can be produced by the following script file:

```
t=0:0.1:6*pi;
x=sqrt(t).*sin(2*t);
y=sqrt(t).*cos(2*t);
z=0.5*t;
plot3(x,y,z,'k','linewidth',1)
grid on
xlabel('x'); ylabel('y'); zlabel('z')
```

When executed the plot that is created is shown in Figure 9-1 below:

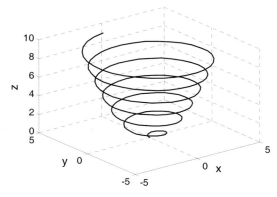

Figure 9-1: A plot of the function $x = \sqrt{t}\sin(2t)$, $y = \sqrt{t}\cos(2t)$, $z = 0.5t$ **for** $0 \le t \le 6\pi$.

9.2 MESH AND SURFACE PLOTS

Mesh and surface plots are three-dimensional plots that are used for plotting functions of the form $z = f(x, y)$ where the x and y are the independent variables and z is the dependent variable. It means that within a given domain the value of z can be calculated for any combination of x and y. Mesh and surface plots are created in three steps. The first step is to create a grid in the x-y plane that covers the domain of the function. The second step is to calculate the value of z at each point of the grid. The third step is to create the plot. The three steps are explained next.

Creating a grid in the x-y plane:

The grid is a set of points in the x-y plane in the domain of the function. The density of the grid (number of points used to define the domain) is defined by the user. Figure 9-2 shows, for example, a grid in the domain $-1 \leq x \leq 3$ and $1 \leq y \leq 4$.

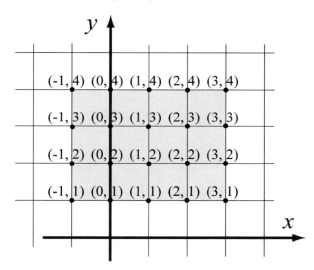

Figure 9-2: A grid in the x-y plane for the domain $-1 \leq x \leq 3$ and $1 \leq y \leq 4$ with spacing of 1.

In this grid the distance between the grid points is one unit. The points of the grid can be defined by two matrices X and Y. Matrix X has the x coordinates of all the points, and matrix Y has the y coordinates of all the points:

$$X = \begin{bmatrix} -1 & 0 & 1 & 2 & 3 \\ -1 & 0 & 1 & 2 & 3 \\ -1 & 0 & 1 & 2 & 3 \\ -1 & 0 & 1 & 2 & 3 \end{bmatrix} \quad \text{and} \quad Y = \begin{bmatrix} 4 & 4 & 4 & 4 & 4 \\ 3 & 3 & 3 & 3 & 3 \\ 2 & 2 & 2 & 2 & 2 \\ 1 & 1 & 1 & 1 & 1 \end{bmatrix}$$

The X matrix is made from identical rows since in each row of the grid the points have the same x coordinate. In the same way the Y matrix is made from identical columns since in each column of the grid the y coordinate of the points is the same.

MATLAB has a built-in function, called `meshgrid`, that can be used for creating the X and Y matrices. The form of the `meshgrid` function is:

$$[X,Y] = \text{meshgrid}(x,y)$$

X is the matrix of the x coordinates of the grid points.
Y is the matrix of the y coordinates of the grid points.

x is a vector that divides the domain of x.
y is a vector that divides the domain of y.

In the vectors x and y the first and the last elements are the respective boundaries of the domain. The density of the grid is determined by the number of elements in the vectors. For example, the mesh matrices X and Y that correspond to the grid in Figure 9-2 can be created with the `meshgrid` command by:

```
>> x=-1:3;
>> y=1:4;
>> [X,Y]=meshgrid(x,y)
X =
      -1       0       1       2       3
      -1       0       1       2       3
      -1       0       1       2       3
      -1       0       1       2       3
Y =
       1       1       1       1       1
       2       2       2       2       2
       3       3       3       3       3
       4       4       4       4       4
```

Once the grid matrices exist, they can be used for calculating the value of z at each grid point.

Calculating the value of z at each point of the grid:

The value of z at each point is calculated by using element-by-element calculations in the same way it is used with vectors. When the independent variables x and y are matrices (must be of the same size) the calculated dependent variable is also a matrix of the same size. The value of z at each address is calculated from the corresponding values of x and y. For example, if z is given by:

$$z = \frac{xy^2}{x^2 + y^2}$$

the value of z at each point of the grid above is calculated by:

```
>> Z = X.*Y.^2./(X.^2 + Y.^2)
```

```
Z =
    -0.5000         0    0.5000    0.4000    0.3000
    -0.8000         0    0.8000    1.0000    0.9231
    -0.9000         0    0.9000    1.3846    1.5000
    -0.9412         0    0.9412    1.6000    1.9200
```

Once the three matrices exist, they can be used to plot mesh or surface plots.

Making mesh and surface plots:

A mesh or a surface plot is created with the `mesh` or `surf` commands, which have the forms:

$$\boxed{\texttt{mesh(X,Y,Z)}} \qquad\qquad \boxed{\texttt{surf(X,Y,Z)}}$$

where X, Y, are matrices with the coordinates of the grid and Z is a matrix with the value of z at the grid points. The mesh plot is made of lines that connect the points. In the surface plot, areas within the mesh lines are colored.

As an example, the following script file contains a complete program that creates the grid and then makes a mesh (or surface) plot of the function $z = \dfrac{xy^2}{x^2 + y^2}$ over the domain $-1 \le x \le 3$ and $1 \le y \le 4$.

```
x=-1:0.1:3;
y=1:0.1:4;
[X,Y]=meshgrid(x,y);
Z=X.*Y.^2./(X.^2+Y.^2);
mesh(X,Y,Z)                    Type surf(X,Y,Z) for surface plot.
xlabel('x'); ylabel('y'); zlabel('z')
```

Note that in the program above the vectors x and y have a much smaller spacing than the spacing earlier in the section. The smaller spacing creates a denser grid. The figures created by the program are:

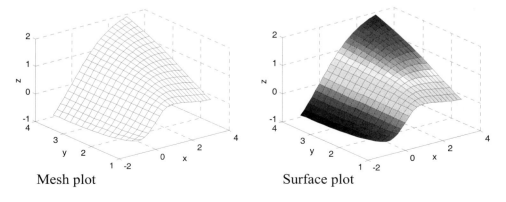

Mesh plot   Surface plot

Additional comments on the mesh command:

- The plots that are created have colors that vary according to the magnitude of z. The variation in color adds to the three-dimensional visualization of the plots. The color can be changed to be a constant either by using the Plot Editor in the Figure Window (select the edit arrow, click on the figure to open the Property Editor Window, then change the color in the Mesh Properties list), or by using the `colormap(C)` command. In this command C is a three-element vector in which the first, second, and third elements specify the intensity of Red, Green, and Blue (RGB) colors, respectively. Each element can be a number between 0 (minimum intensity) and 1 (maximum intensity). Some typical colors are:

  C = [0 0 0]   black   C = [1 0 0]   red   C = [0 1 0]   green
  C = [0 0 1]   blue   C = [1 1 0]   yellow   C = [1 0 1]   magenta
  C = [0.5 0.5 0.5]   gray

- When the mesh command executes, the grid is on by default. The grid can be turned off with the `grid off` command.

- A box can be drawn around the plot with the `box on` command.

- The `mesh` and `surf` commands can also be used with the form `mesh(Z)` and `surf(z)`. In this case the values of Z are plotted as a function of their addresses in the matrix. The row number is on the x axis and the column number is on the y axis.

There are several additional plotting commands that are similar to the `mesh` and `surf` commands that create plots with different features. Table 9-1 shows a summary of mesh and surface plotting commands. All the examples in the table are plots of the function $z = 1.8^{-1.5\sqrt{x^2+y^2}}\sin(x)\cos(0.5y)$ over the domain $-3 \le x \le 3$ and $-3 \le y \le 3$.

Table 9-1: Mesh and surface plots

Plot type	Example of plot	Program

Mesh Plot
 	
Function format:
mesh(X,Y,Z)
		```x=-3:0.25:3;```
		```y=-3:0.25:3;```
		```[X,Y] = meshgrid(x,y);```
		```Z=1.8.^(-1.5*sqrt(X.^2+Y.^2)).*cos(0.5*Y).*sin(X);```
		```mesh(X,Y,Z)```
		```xlabel('x'); ylabel('y')```
		```zlabel('z')```

Table 9-1: Mesh and surface plots (Continued)

Plot type	Example of plot	Program

Surface Plot
 	
Function format:
`surf(X,Y,Z)`
		```x=-3:0.25:3;```
		```y=-3:0.25:3;```
		```[X,Y]=meshgrid(x,y);```
		```Z=1.8.^(-1.5*sqrt(X.^2+```
		```Y.^2)).*cos(0.5*Y).*sin(X);```
		```surf(X,Y,Z)```
		```xlabel('x'); ylabel('y')```
		```zlabel('z')```

Mesh Curtain Plot (draws a curtain around the mesh)
 	
Function format:
`meshz(X,Y,Z)`
		```x=-3:0.25:3;```
		```y=-3:0.25:3;```
		```[X,Y]=meshgrid(x,y);```
		```Z=1.8.^(-1.5*sqrt(X.^2+```
		```Y.^2)).*cos(0.5*Y).*sin(X);```
		```meshz(X,Y,Z)```
		```xlabel('x'); ylabel('y')```
		```zlabel('z')```

Mesh and Contour Plot (draws a contour plot beneath the mesh)
 	
Function format:
`meshc(X,Y,Z)`
		```x=-3:0.25:3;```
		```y=-3:0.25:3;```
		```[X,Y]=meshgrid(x,y);```
		```Z=1.8.^(-1.5*sqrt(X.^2+```
		```Y.^2)).*cos(0.5*Y).*sin(X);```
		```meshc(X,Y,Z)```
		```xlabel('x'); ylabel('y')```
		```zlabel('z')```

Surface and Contour Plot (draws a contour plot beneath the surface)
 	
Function format:
`surfc(X,Y,Z)`
		```x=-3:0.25:3;```
		```y=-3:0.25:3;```
		```[X,Y]=meshgrid(x,y);```
		```Z=1.8.^(-1.5*sqrt(X.^2+```
		```Y.^2)).*cos(0.5*Y).*sin(X);```
		```surfc(X,Y,Z)```
		```xlabel('x'); ylabel('y')```
		```zlabel('z')```

Table 9-1: Mesh and surface plots (Continued)

Plot type	Example of plot	Program

Surface Plot with Lighting
 	
Function format:
`surfl(X,Y,Z)`
		```x=-3:0.25:3;```
		```y=-3:0.25:3;```
		```[X,Y]=meshgrid(x,y);```
		```Z=1.8.^(-1.5*sqrt(X.^2+```
		```Y.^2)).*cos(0.5*Y).*sin(X);```
		```surfl(X,Y,Z)```
		```xlabel('x'); ylabel('y')```
		```zlabel('z')```

Waterfall Plot (draws a mesh in one direction only)
 	
Function format:
`water-fall(X,Y,Z)`
		```x=-3:0.25:3;```
		```y=-3:0.25:3;```
		```[X,Y] = meshgrid(x,y);```
		```Z=1.8.^(-1.5*sqrt(X.^2+```
		```Y.^2)).*cos(0.5*Y).*sin(X);```
		```waterfall(X,Y,Z)```
		```xlabel('x'); ylabel('y')```
		```zlabel('z')```

3-D Contour Plot
 	
Function format:
`contour3(X,Y,Z,n)`
 	
n is the number of contour levels (optional)
		```x=-3:0.25:3;```
		```y=-3:0.25:3;```
		```[X,Y]=meshgrid(x,y);```
		```Z=1.8.^(-1.5*sqrt(X.^2+```
		```Y.^2)).*cos(0.5*Y).*sin(X);```
		```contour3(X,Y,Z,15)```
		```xlabel('x'); ylabel('y')```
		```zlabel('z')```

2-D Contour Plot (draws projections of contour levels on the x-y plane)
Function format:
`contour (X,Y,Z,n)`
n is the number of contour levels (optional)
		```x=-3:0.25:3;```
		```y=-3:0.25:3;```
		```[X,Y]=meshgrid(x,y);```
		```Z=1.8.^(-1.5*sqrt(X.^2+```
		```Y.^2)).*cos(0.5*Y).*sin(X);```
		```contour(X,Y,Z,15)```
		```xlabel('x'); ylabel('y')```
		```zlabel('z')```

9.3 PLOTS WITH SPECIAL GRAPHICS

MATLAB has additional functions for creating various types of special three-dimensional plots. A complete list can be found in the Help Window under Plotting and Data Visualization. Several of these 3-D plots are presented in Table 9-2.

Table 9-2: Specialized 3-D plots

Plot type	Example of plot	Program

Plot a Sphere
 	
Function format:
`sphere`
Returns the x, y, and z coordinates of a unit sphere with 20 faces.
`sphere(n)`
Same as above with n faces.
		`sphere`
		
		or:
		
		`[X,Y,Z]=sphere(20);`
		`surf(X,Y,Z)`

Plot a Cylinder
 	
Function format:
`[X,Y,Z]=`
`cylinder(r)`
Returns the x, y, and z coordinates of cylinder with profile r.
		`t=linspace(0,pi,20);`
		`r=1+sin(t);`
		`[X,Y,Z]=cylinder(r);`
		`surf(X,Y,Z)`
		`axis square`

3-D Bar Plot
 	
Function format:
`bar3(Y)`
Each element in `Y` is one bar. Columns are grouped together.
		`Y=[1 6.5 7; 2 6 7; 3`
		`5.5 7; 4 5 7; 3 4 7;`
		`2 3 7; 1 2 7];`
		
		`bar3(Y)`

Table 9-2: Specialized 3-D plots (Continued)

Plot type	Example of plot	Program

3-D Stem Plot (draws sequential points with markers and vertical lines from the *x-y* plane)
 	
Function format:
stem3(X,Y,Z)
		```t=0:0.2:10;```
		```x=t;```
		```y=sin(t);```
		```z=t.^1.5;```
		```stem3(x,y,z,'fill')```
		```grid on```
		```xlabel('x');```
		```ylabel('y')```
		```zlabel('z')```

3-D Scatter Plot
 	
Function format:
scatter3(X, Y,Z)
		```t=0:0.4:10;```
		```x=t;```
		```y=sin(t);```
		```z=t.^1.5;```
		```scatter3(x,y,z,'filled')```
		```grid on```
		```colormap([0.1 0.1 0.1])```
		```xlabel('x');```
		```ylabel('y')```
		```zlabel('z')```

3-D Pie Plot
 	
Function format:
pie3(X, explode)
		```X=[5 9 14 20];```
		```explode=[0 0 1 0];```
		```pie3(X,explode)```
		
		explode is a vector (same length as X) of 0's and 1's. 1 offsets the slice from the center.

The examples in the table do not show all the options available with each plot type. More details for each type of plot can be obtained in the Help Window, or by typing help command_name in the Command Window.

9.4 THE `view` *COMMAND*

The `view` command controls the direction from which the plot is viewed. This is done by specifying a direction in terms of azimuth and elevation angles, as seen in Figure 9-3, or by defining a point in space from which the plot is viewed. To set the viewing angle of the plot, the `view` command has the form:

> `view(az,el) or view([az,el])`

- `az` is the azimuth, which is an angle (in degrees) in the *x-y* plane measured relative to the negative *y* axis direction and defined as positive in the counterclockwise direction.

- `el` is the angle of elevation (in degrees) from the *x-y* plane. A positive value corresponds to opening an angle in the direction of the *z* axis.

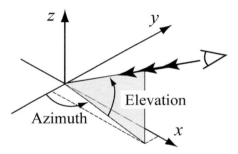

Figure 9-3: Azimuth and Elevation angles.

- The default view angles are $az = -37.5^o$, and $el = 30^o$.

For example, the surface plot from Table 9-1 is plotted again in Figure 9-4 below with a viewing angles of $az = 20^o$, and $el = 35^o$.

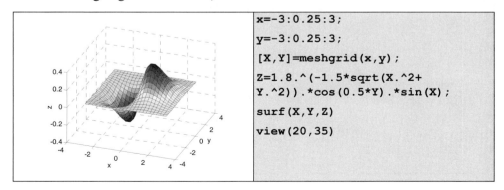

```
x=-3:0.25:3;
y=-3:0.25:3;
[X,Y]=meshgrid(x,y);
Z=1.8.^(-1.5*sqrt(X.^2+
Y.^2)).*cos(0.5*Y).*sin(X);
surf(X,Y,Z)
view(20,35)
```

Figure 9-4: A surface plot of the function $z = 1.8^{-1.5\sqrt{x^2+y^2}} \sin(x)\cos(0.5y)$ **with viewing angles of** $az = 20^o$, **and** $el = 35^o$.

- By choosing appropriate azimuth and elevation angles the `view` command can be used to plot projections of 3-D plots on various planes according to the following table:

Projection plane	*az* value	*el* value
x-y (top view)	0	90
x-z (side view)	0	0
y-z (side view)	90	0

An example of a top view is shown next. The figure shows the top view of the function that is plotted in Figure 9-1.

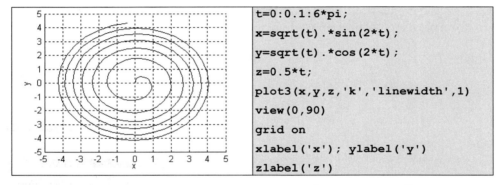

```
t=0:0.1:6*pi;
x=sqrt(t).*sin(2*t);
y=sqrt(t).*cos(2*t);
z=0.5*t;
plot3(x,y,z,'k','linewidth',1)
view(0,90)
grid on
xlabel('x'); ylabel('y')
zlabel('z')
```

Figure 9-5: A top view plot of the function $x = \sqrt{t}\sin(2t)$, $y = \sqrt{t}\cos(2t)$,

$z = 0.5t$ **for** $0 \le t \le 6\pi$.

Examples of projections onto the *x-z* and *y-z* planes are shown next in Figures 9-6 and 9-7, respectively. The figures show mesh plot projections of the function plotted in Table 9-1.

```
x=-3:0.25:3;
y=-3:0.25:3;
[X,Y]=meshgrid(x,y);
Z=1.8.^(-1.5*sqrt(X.^2+
Y.^2)).*cos(0.5*Y).*sin(X);
mesh(X,Y,Z)
view(0,0)
```

Figure 9-6: Projections on the *x-z* plane of the function

$z = 1.8^{-1.5\sqrt{x^2+y^2}}\sin(x)\cos(0.5y)$.

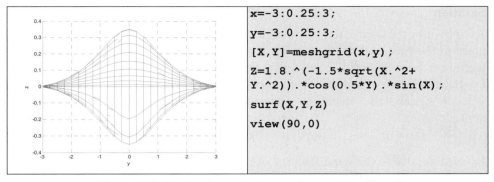

```
x=-3:0.25:3;
y=-3:0.25:3;
[X,Y]=meshgrid(x,y);
Z=1.8.^(-1.5*sqrt(X.^2+
Y.^2)).*cos(0.5*Y).*sin(X);
surf(X,Y,Z)
view(90,0)
```

Figure 9-7: Projections on the *y-z* planes of the function

$$z = 1.8^{-1.5\sqrt{x^2+y^2}}\sin(x)\cos(0.5y).$$

- The view command can also set a default view:

  view(2)   sets the default to the top view which is a projection onto the *x-y* plane with $az = 0°$, and $el = 90°$.

  view(3)   sets the default to the standard 3-D view with $az = -37.5°$, and $el = 30°$.

- The viewing direction can also be set by selecting a point in space from which the plot is viewed. In this case the view command has the form view([x,y,z]), where x, y, and z are the coordinates of the point. The direction is determined by the direction from the specified point to the origin of the coordinate system and is independent of the distance. This means that the view is the same with point [6, 6, 6] as with point [10, 10, 10]. Top view can be set up with [0, 0, 1]. A side view of the *x-z* plane from the negative *y* direction can be set with [0, −1, 0], and so on.

9.5 *EXAMPLES OF MATLAB APPLICATIONS*

Sample Problem 9-1: 3-D projectile trajectory

A projectile is fired with an initial velocity of 250 m/s at an angle of $\theta = 65°$ relative to the ground. The projectile is aimed directly North. Because of a strong wind blowing to the West, the projectile also moves in this direction at a constant speed of 30 m/s. Determine and plot the trajectory of the projectile until it hits the ground. For comparison, plot also (in the same figure) the trajectory that the projectile would have had if there was no wind.

Solution

As shown in the figure, the coordinate system is set up such that the x and y axes point to the East and North directions, respectively. Then, the motion of the projectile can be analyzed by considering the vertical direction z and the two horizontal components x and y. Since the projectile is fired directly North, the initial velocity v_0 can be resolved into a horizontal y component and a vertical z component:

$$v_{0y} = v_0\cos(\theta) \quad \text{and} \quad v_{0z} = v_0\sin(\theta)$$

In addition, due to the wind the projectile has a constant velocity in the negative x direction, $v_x = -30$ m/s.

The initial position of the projectile (x_0, y_0, z_0) is at point (3000, 0, 0). In the vertical direction the velocity and position of the projectile are given by:

$$v_z = v_{0z} - gt \quad \text{and} \quad z = z_0 + v_{0z}t - \frac{1}{2}gt^2$$

The time it takes the projectile to reach the highest point $(v_z = 0)$ is $t_{hmax} = \dfrac{v_{0z}}{g}$. The total flying time is twice this time, $t_{tot} = 2t_{hmax}$. In the horizontal direction the velocity is constant (both in the x and y directions), and the position of the projectile is given by:

$$x = x_0 + v_xt \quad \text{and} \quad y = y_0 + v_{0y}t$$

The following MATLAB program written in a script file solves the problem by following the equations above.

```
v0=250; g=9.81; theta=65;
x0=3000; vx=-30;
v0z=v0*sin(theta*pi/180);
v0y=v0*cos(theta*pi/180);
t=2*v0z/g;
tplot=linspace(0,t,100);          Creating a time vector with 100 elements.
z=v0z*tplot-0.5*g*tplot.^2;
y=v0y*tplot;                       Calculating the x, y and z coordinates
                                   of the projectile at each time.
x=x0+vx*tplot;
xnowind(1:length(y))=x0;          Constant x coordinate when no wind.
plot3(x,y,z,'k-',xnowind,y,z,'k--')          Two 3-D line plots.
grid on
axis([0 6000 0 6000 0 2500])
xlabel('x (m)'); ylabel('y (m)'); zlabel('z (m)')
```

The figure generated by the program is shown below.

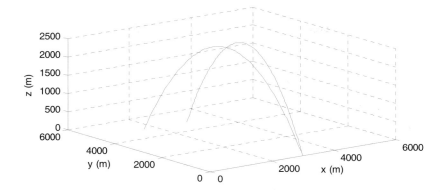

=====

Sample Problem 9-2: Electric potential of two point charges

The electric potential V around a charged particle is given by:

$$V = \frac{1}{4\pi\varepsilon_0}\frac{q}{r}$$

where $\varepsilon_0 = 8.8541878 \times 10^{-12}\dfrac{C}{N \cdot m^2}$ is the permittivity constant, q is the magnitude of the charge in Coulombs, and r is the distance from the particle in meters. The electric field of two or more particles is calculated by using superposition. For example, the electric potential at a point due to two particles is given by:

$$V = \frac{1}{4\pi\varepsilon_0}\left(\frac{q_1}{r_1} + \frac{q_2}{r_2}\right)$$

where q_1, q_2, r_1, and r_2 are the charges of the particles and the distance from the point to the corresponding particle, respectively.

Two particles with a charge of $q_1 = 2 \times 10^{-10}$ C and $q_2 = 3 \times 10^{-10}$ C are positioned in the x-y plane at points (0.25, 0, 0) and (−0.25, 0, 0), respectively as shown. Calculate and plot the electric potential due to the two particles at points in the x-y plane that are located in the domain $-0.2 \le x \le 0.2$ and $-0.2 \le y \le 0.2$ (the units in the x-y plane are in meters). Make the plot such that the x-y plane is the plane of the points, and the z axis is the magnitude of the electric potential.

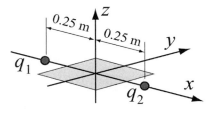

Solution

The problem is solved following these steps:

a) A grid is created in the *x-y* plane with the domain $-0.2 \le x \le 0.2$ and $-0.2 \le y \le 0.2$.
b) The distance from each grid point to each of the charges is calculated.
c) The electric potential at each point is calculated.
d) The electric potential is plotted.

The following is a program in a script file that solves the problem.

```
eps0=8.85e-12; q1=2e-10; q2=3e-10;
k=1/(4*pi*eps0);
x=-0.2:0.01:0.2;
y=-0.2:0.01:0.2;
[X,Y]=meshgrid(x,y);          Creating a grid in the x-y plane.
r1=sqrt((X+0.25).^2+Y.^2);    Calculating the distance r₁ for each grid point.
r2=sqrt((X-0.25).^2+Y.^2);    Calculating the distance r₂ for each grid point.
V=k*(q1./r1+q2./r2);          Calculating the electric potential V at each grid point.
mesh(X,Y,V)
xlabel('x (m)'); ylabel('y (m)'); zlabel('V (V)')
```

The plot generated when the program runs is:

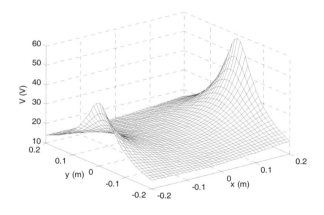

Sample Problem 9-3:   Heat conduction in a square plate

Three sides of a rectangular plate ($a = 5$ m, $b = 4$ m) are kept at a temperature of 0°C and one side is kept at a temperature of $T_1 = 80\,^\circ$C, as shown in the figure. Determine and plot the temperature distribution $T(x,y)$ in the plate.

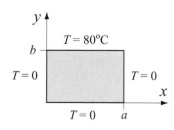

Solution

The temperature distribution, $T(x,y)$ in the plate can be determined by solving the two-dimensional heat equation. For the given boundary conditions $T(x,y)$ can be expressed analytically by Fourier series (Erwin Kreyszig, "Advanced Engineering Mathematics," John Wiley and Sons, 1993):

$$T(x, y) = \frac{4T_1}{\pi} \sum_{n=1}^{\infty} \frac{\sin\left[(2n-1)\frac{\pi x}{a}\right]\sinh\left[(2n-1)\frac{\pi y}{a}\right]}{(2n-1)} \frac{}{\sinh\left[(2n-1)\frac{\pi b}{a}\right]}$$

A program in a script file that solves the problem is listed below. The program follows these steps:

a) Create an X, Y grid in the domain $0 \le x \le a$ and $0 \le y \le b$. The length of the plate, a, is divided into 20 segments, and the width of the plate, b, is divided into 16 segments.

b) Calculate the temperature at each point of the mesh. The calculations are done point by point using a double loop. At each point the temperature is determined by adding k terms of the Fourier series.

c) Make a surface plot of T.

```
a=5; b=4; na=20; nb=16; k=5; T0=80;
clear T
x=linspace(0,a,na);
y=linspace(0,b,nb);
[X,Y]=meshgrid(x,y);                    Creating a grid in the x-y plane.
for i=1:nb                              First loop, i is the index of the grid's row.
    for j=1:na                          Second loop, j is the index of the grid's column.
        T(i,j)=0;
        for n=1:k                       Third loop, n is the nth term of the Fourier
            ns=2*n-1;                    series, k is the number of terms.
    T(i,j)=T(i,j)+sin(ns*pi*X(i,j)/a).*sinh(ns*pi*Y(i,j)/
a)/(sinh(ns*pi*b/a)*ns);
        end
        T(i,j) = T(i,j)*4*T0/pi;
    end
end
mesh(X,Y,T)
xlabel('x (m)'); ylabel('y (m)'); zlabel('T ( ^oC)')
```

The program was executed twice, first using 5 terms ($k = 5$) in the Fourier series to calculate the temperature at each point, and then with $k = 50$. The mesh plots created in each execution are shown in the figures below. The temperature should be uniformly 80°C at $y = 4$ m. Note the effect of the number of terms (k) on the accuracy at $y = 4$ m.

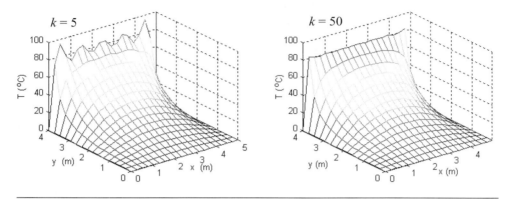

9.6 PROBLEMS

1. The position of a moving particle as a function of time is given by:

    $$x = (1 + 0.1t)\cos(t)$$
    $$y = x = (1 + 0.1t)\sin(t)$$
    $$z = 0.2\sqrt{t}$$

    Plot the position of the particle for $0 \le t \le 30$.

2. An elliptical staircase of height h can be modeled by the parametric equations:

    $$x = r\cos(t)$$
    $$y = r\sin(t)$$
    $$z = \frac{ht}{2\pi n}$$

    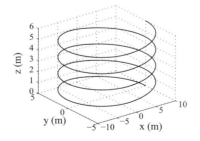

    where $r = \dfrac{b}{\sqrt{1 - \varepsilon \cos^2(t)}}$. ε is the eccen-

    tricity of the ellipse defined by $\varepsilon = \sqrt{1 - \dfrac{b^2}{a^2}}$. a and b are the semimajor and semiminor axes of the ellipse, and n is the number of revolutions that the staircase makes. Make a 3-D plot of an elliptical staircase with $a = 10$ m, $b = 5$ m, $h = 6$ m, and $n = 4$. (Create a vector t for the domain 0 to $2\pi n$, and use the plot3 command.)

3. The ladder of a fire truck can be elevated (increase of angle ϕ), rotated about the z axis (increase of angle θ), and extended (increase of r). Initially the ladder rests on the truck, $\phi = 0$, $\theta = 0$, and $r = 8$ m. Then the ladder is moved to a new position by raising the ladder at a rate of 5 deg/s, rotating at a rate of 8 deg/s, and extending the ladder at a rate of 0.6 m/s. Determine and plot the position of the tip of the ladder for 10 seconds.

4. Make a 3-D surface plot and a 3-D contour plot of the function $z = 4\sqrt{\dfrac{x^2}{2} + \dfrac{y^2}{2} + 1}$ in the domain $-2 \le x \le 2$ and $-2 \le y \le 2$.

5. An anti-symmetric cross-ply composite laminate has two layers where the fibers are aligned perpendicular to one another. A laminate of this type will deform into a saddle shape due to residual thermal stresses as described by the equation:

$$w = k(x^2 - y^2)$$

where x and y are the in-plane coordinates, w is the out-of-plane deflection and k is the curvature (a complicated function of material properties and geometry). Make a surface plot showing the deflection of a six-inch square plate ($-3 \le x \le 3$ in., $-3 \le y \le 3$ in.) assuming $k = 0.01$ in^{-1}.

6. Make a 3-D surface plot of the function $z = \sin\left(\dfrac{2\pi x}{60}\right)\sin\left(\dfrac{3\pi y}{60}\right)$ in the domain $0 \le x \le 60$ and $0 \le y \le 60$.

7. The van der Waals equation gives a relationship between the pressure p (in atm.), volume V, (in L), and temperature T (in K) for a real gas:

$$P = \frac{nRT}{V - b} - \frac{n^2 a}{V^2}$$

where n is the number of moles, $R = 0.08206$ (L atm)/(mole K) is the gas constant, and a (in L^2 atm/mole^2), and b (in L/mole) are material constants.

Consider 1.5 moles of nitrogen ($a = 1.39$ L^{2}atm/mole^2, $b = 0.03913$ L/mole). Make a 3-D plot that shows the variation of pressure (dependent variable, z axis) with volume (independent variable, x axis) and temperature (independent variable, y axis). The domains for the volume and temperature are: $0.3 \le V \le 1.2$ L, and $273 \le T \le 473$ K.

8. Molecules of a gas in a container are moving around at different speeds. Maxwell's speed distribution law gives the probability distribution $P(v)$ as a function of temperature and speed:

$$P(v) = 4\pi \left(\frac{M}{2\pi RT} \right)^{3/2} v^2 e^{(-Mv^2)/(2RT)}$$

where M is the molar mass of the gas in kg/mol, $R = 8.31$ J/mol-K is the gas constant, T is the temperature in degrees K, and v is the molecules speed in m/s.

   Make a 3-D plot of $P(v)$ as a function of v and T for $0 \le v \le 1000$ m/s and $70 \le T \le 320$ K for oxygen (molar mass 0.032 kg/mol).

9. The heat index (sometimes called apparent temperature) is a measure of how the temperature actually feels when the effect of relative humidity is added. One formula that is used for calculating the heat index is (www.noaa.gov):

$$HI = -42.379 + 2.04901523T + 10.14333127R_H - 0.22475541\,T\,R_H -$$

$$- 6.83783 \times 10^{-3}T^2 - 5.481717 \times 10^{-2}\,R_H^2 + 1.22874 \times 10^{-3}\,T^2\,R_H +$$

$$+ 8.5282 \times 10^{-4}\,T\,R_H^2 - 1.99 \times 10^{-6}T^2\,R_H^2$$

where HI is the heat index, T is the temperature in degrees F, and R_H in %.

   Make a 3-D plot of HI as a function of T and R_H for $80° \le T \le 105°$F and $30 \le R_H \le 90$ %.

10. An RLC circuit with an alternating voltage source is shown. The source voltage v_s is given by $v_s = v_m \sin(\omega_d t)$ where $\omega_d = 2\pi f_d$ in which f_d is the driving frequency. The amplitude of the current, I, in this circuit is given by:

$$I = \frac{v_m}{\sqrt{R^2 + (\omega_d L - 1/(\omega_d C))^2}}$$

where R and C are the resistance of the resistor and capacitance of the capacitor, respectively. For the circuit in the figure $C = 15 \times 10^{-6}$ F, $L = 240 \times 10^{-3}$ H, and $v_m = 24$ V.

   a) Make a 3-D plot of I (z axis) as a function of ω_d (x axis) for $60 \le f \le 110$ Hz, and a function of R (y axis) for $10 \le R \le 40\ \Omega$.

   b) Make a plot that is a projection on the x-z plane. Estimate from this plot the natural frequency of the circuit (the frequency at which I is maximum). Compare the estimate with the calculated value of $1/(2\pi\sqrt{LC})$.

11. The normal stress σ_{xx} at point (y, z) in the cross section of a rectangular beam, due to the applied force F at a point (y_F, z_F), is given by:

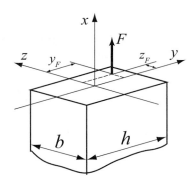

$$\sigma_{xx} = \frac{F}{A} + \frac{F z_F z}{I_{yy}} + \frac{F y_F y}{I_{zz}}$$

where I_{zz} and I_{yy} are the area moments of inertia defined by:

$$I_{zz} = \frac{1}{12}bh^3 \text{ and } I_{yy} = \frac{1}{12}hb^3 .$$

Determine and plot the normal stress in the cross-sectional area shown in the figure, given that: $h = 40$ mm, $b = 30$ mm, $y_F = -15$mm, $z_F = -10$mm, and $F = -250000$ N. Plot the coordinates y and z in the horizontal plane, and the normal stress in the vertical direction.

12. A defect in a crystal lattice where a row of atoms is missing is called an edge dislocation. The stress field around an edge dislocation is given by:

$$\sigma_{xx} = \frac{-Gb}{2\pi(1-v)}\frac{y(3x^2 + y^2)}{(x^2 + y^2)^2}$$

$$\sigma_{yy} = \frac{Gb}{2\pi(1-v)}\frac{y(x^2 - y^2)}{(x^2 + y^2)^2}$$

$$\tau_{xy} = \frac{Gb}{2\pi(1-v)}\frac{x(x^2 - y^2)}{(x^2 + y^2)^2}$$

Edge dislocation

Plot domain

where G is the shear modulus, b is the Burgers vector, and v is Poisson's ratio. Plot the stress components (each in a separate figure) due to an edge dislocation in aluminum for which $G = 27.7 \times 10^9$ Pa, $b = 0.286 \times 10^{-9}$m, and $v = 0.334$. Plot the stresses in the domain $-5 \times 10^{-9} \le x \le 5 \times 10^{-9}$ m and $-5 \times 10^{-9} \le y \le -1 \times 10^{-9}$m. Plot the coordinates x and y in the horizontal plane, and the stresses in the vertical direction.

13. The current I flowing through a semiconductor diode is given by:

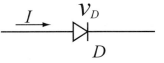

$$I = I_S\left(e^{\frac{qv_D}{kT}} - 1\right)$$

where $I_S = 10^{-12}$ A is the saturation current,

$q = 1.6 \times 10^{-12}$ coulombs is the elementary charge value, $k = 1.38 \times 10^{-23}$ joule/K is Boltzmann's constant, v_D is the voltage drop across the diode, and T is the temperature in Kelvins. Make a 3-D plot of I (z axis) versus v_D (x axis) for $0 \leq v_D \leq 0.4$, and versus T (y axis) for $290 \leq T \leq 320\,\mathrm{K}$.

14. The equation for the streamlines for uniform flow over a cylinder is:

$$\psi(x, y) = y - \frac{y}{x^2 + y^2}$$

where ψ is the stream function. For example, if $\psi = 0$ then $y = 0$. Since the equation is satisfied for all x, the x-axis is the zero ($\psi = 0$) streamline. Observe also that the collection of points where $x^2 + y^2 = 1$ is also a stream-line. Thus, the stream function above is for a cylinder of radius 1. Make a 2-D contour plot of the streamlines around a one inch radius cylinder. Set up the domain for x and y to range between -3 and 3. Use 100 for the number of contour levels. Add to the figure a plot of a circle with radius of 1. Note that MATLAB also plots streamlines inside the cylinder. This is a mathematical artifact.

15. The Verhulst model, given in the following equation, describes the growth of a population that is limited by various factors such as overcrowding, lack of resources etc.:

$$N(t) = \frac{N_\infty}{1 + \left(\frac{N_\infty}{N_0} - 1\right)e^{-rt}}$$

where $N(t)$ is the number of individuals in the population, N_0 is the initial population size, N_∞ is the maximum population size possible due to the various limiting factors, and r is a rate constant. Make a surface plot of $N(t)$ versus t and N_∞ assuming $r = 0.1\,\mathrm{s}^{-1}$, and $N_0 = 10$. Let t vary between 0 and 100 and N_∞ between 100 and 1000.

Chapter 10

Applications in Numerical Analysis

Numerical methods are commonly used for solving mathematical problems that are formulated in science and engineering where it is difficult or even impossible to obtain exact solutions. MATLAB has a large library of functions for numerically solving a wide variety of mathematical problems. This chapter explains how to use a number of the most frequently used of these functions. It should be pointed out here that the purpose of this book is to show users how to use MATLAB. Some general information on the numerical methods is given, but the details, which can be found in books on numerical analysis, are not included.

The following topics are presented in this chapter: solving an equation with one unknown, finding a minimum or a maximum of a function, numerical integration, and solving a first-order ordinary differential equation.

10.1 SOLVING AN EQUATION WITH ONE VARIABLE

An equation with one variable can be written in the form $f(x) = 0$. The solution is the value of x where the function crosses the x axis (the value of the function is zero), which means that the function changes sign at x. An exact solution is a value of x for which the value of the function is exactly zero. If such a value does not exist or is difficult to determine, a numerical solution can be determined by finding an x that is very close to the point where the function changes its sign (crosses the x-axis). This is done by the iterative process where in each iteration the computer determines a value of x that is closer to the solution. The iterations stop when the difference in x between two iterations is smaller than some measure. In general, a function can have none, one, several, or infinite number of solutions.

In MATLAB a zero of a function can be determined with the command (built-in function) `fzero` that has the form:

Solution   The function to   A value of x near to where the
                 be solved.   function crosses the axis.

The built-in function `fzero` is a MATLAB function function (see Section 6.9) which means that it accepts another function (the function to be solved) as an input argument.

Additional details on the arguments of `fzero`:

- `x` is the solution, which is a scalar.

- `function` is the function to be solved. It can be entered in several different ways:
  1. The simplest way is to enter the mathematical expression as a string.
  2. The function is first created as a user-defined function in a function file and then the function handle is entered (see Section 6.9.1).
  3. The function is first created as an anonymous function (see Section 6.8.1) and then the name of the anonymous function (which is the name of the handle) is entered (see Section 6.9.1).

  (As explained in Section 6.9.2, it is also possible to pass a user-defined function and an inline function into a function function by using its name. However, function handles are more efficient and easier to use, and should be the preferred method.)

- The function has to be written in a standard form. For example, if the function to be solved is $xe^{-x} = 0.2$, it has to be written as $f(x) = xe^{-x} - 0.2 = 0$. If this function is entered into the `fzero` command as a string, it is typed as: `'x*exp(-x)-0.2'`.

- When a function is entered as an expression (string), it cannot include predefined variables. For example, if the function to be entered is $f(x) = xe^{-x} - 0.2$, it is not possible to define `b=0.2` and then enter `'x*exp(-x)-b'`.

- `x0` can be a scalar or a two-element vector. If it is entered as a scalar, it has to be a value of x near the point where the function crosses the x axis. If `x0` is entered as a vector, the two elements have to be points on opposite sides of the solution such that $f(x0(1))$ has a different sign than $f(x0(2))$. When a function has more than one solution, each solution can be determined separately by using the `fzero` function and entering values for `x0` that are near each of the solutions.

- A good way to find approximately where a function has a solution is to make a plot of the function. In many applications in science and engineering the domain of the solution can be estimated. Often when a function has more than one solution only one of the solutions will have a physical meaning.

Sample Problem 10-1: Solving a nonlinear equation

Determine the solution of the equation $xe^{-x} = 0.2$.

Solution

The equation is first written in a form of a function: $f(x) = xe^{-x} - 0.2$. A plot of the function, shown on the right, shows that the function has one solution between 0 and 1 and another solution between 2 and 3. The plot is obtained by typing

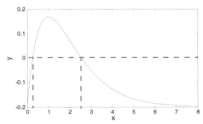

```
>> fplot('x*exp(-x)-0.2',[0 8])
```

in the Command Window. The solutions of the function are found by using the fzero command twice. First, the equation is entered as a string expression, and a value of x0 between 0 and 1, (x0 = 0.7) is used. Second, the equation to be solved is written as an anonymous function, which is then used in fzero with x0 between 2 and 3, (x0 = 2.8). This is shown below:

```
>> x1=fzero('x*exp(-x)-0.2',0.7)        The function is entered as a
x1 =                                    string expression.
    0.2592                              The first solution is 0.2592.
>> F=@(x)x*exp(-x)-0.2
F =                                     Creating an anonymous function.
    @(x)x*exp(-x)-0.2
>> fzero(F,2.8)        Using the name of the anonymous function in fzero.
ans =
    2.5426                              The second solution is 2.5426.
```

Additional comments:

- The fzero command finds zeros of a function only where the function crosses the x-axis. The command does not find a zero at points where the function touches but does not cross the x-axis.

- If a solution cannot be determined, NaN is assigned to x.

- The `fzero` command has additional options (see the Help Window). Two of the more important options are:

  `[x fval]=fzero(function, x0)` assigns the value of the function at x to the variable `fval`.

  `x=fzero(function, x0, optimset('display','iter'))` displays the output of each iteration during the process of finding the solution.

- When the function can be written in the form of a polynomial, the solution, or the roots, can be found with the `roots` command, as is explained in Chapter 8 (Section 8.1.2).

- The `fzero` command can also be used to find the value of x where the function has a specific value. This is done by translating the function up or down. For example, in the function of Sample Problem 10-1 the first value of x where the function is equal to 0.1 can be determined by solving the equation $xe^{-x} - 0.3 = 0$. This is shown below:

```
>> x=fzero('x*exp(-x)-0.3',0.5)

x =
      0.4894
```

10.2 FINDING A MINIMUM OR A MAXIMUM OF A FUNCTION

In many applications there is a need to determine the local minimum or the maximum of a function of the form $y = f(x)$. In calculus the value of x that corresponds to a local minimum or the maximum is determined by finding the zero of the derivative of the function. The value of y is determined by substituting the x into the function. In MATLAB the value of x where a one-variable function $f(x)$ within the interval $x_1 \le x \le x_2$ has a minimum can be determined with the `fmin-bnd` command which has the form:

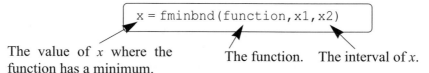

| x = fminbnd(function,x1,x2) |

The value of x where the function has a minimum.   The function.   The interval of x.

- The function can be entered as a string expression, or as a function handle, in the same way as with the `fzero` command. See Section 10.1 for details.

- The value of the function at the minimum can be added to the output by using the option:

  `[x fval]=fminbnd(function,x1,x2)`

  where the value of the function at x is assigned to the variable `fval`.

- Within a given interval, the minimum of a function can either be at one of the end points of the interval, or at a point within the interval where the slope of the function is zero (local minimum). When the `fminbnd` command is executed, MATLAB looks for a local minimum. If a local minimum is found, its value is compared to the value of the function at the end points of the interval. MATLAB returns the point with the actual minimum value of the interval.

For example, consider the function $f(x) = x^3 - 12x^2 + 40.25x - 36.5$ which is plotted in the interval $0 \le x \le 8$ in the figure on the right. It can be observed that there is a local minimum between 5 and 6, and that the absolute minimum is at $x = 0$. Using the `fminbnd` command with the interval $3 \le x \le 8$ to find the location of the local minimum and the value of the function at this point gives:

```
>> [x fval]=fminbnd('x^3-12*x^2+40.25*x-36.5',3,8)

x =
     5.6073
fval =
    -11.8043
```

The local minimum is at $x = 5.6073$. The value of the function at this point is -11.8043.

Notice that the `fminbnd` command gives the local minimum. If the interval is changed to $0 \le x \le 8$, the `fminbnd` gives:

```
>> [x fval]=fminbnd('x^3-12*x^2+40.25*x-36.5',0,8)

x =
     0
fval =
    -36.5000
```

The minimum is at $x = 0$. The value of the function at this point is -36.5.

For this interval the `fminbnd` command gives the absolute minimum which is at the end point $x = 0$.

* The `fminbnd` command can also be used to find the maximum of a function. This is done by multiplying the function by -1 and finding the minimum. For example, the maximum of the function $f(x) = xe^{-x} - 0.2$, (from Sample Problem 10-1) in the interval $0 \le x \le 8$ can be determined by finding the minimum of the function $f(x) = -xe^{-x} + 0.2$ as shown below:

```
>> [x fval]=fminbnd('-x*exp(-x)+0.2',0,8)

x =
     1.0000
fval =
    -0.1679
```

The maximum is at $x = 1.0$. The value of the function at this point is 0.1679.

10.3 *Numerical Integration*

Integration is a common mathematical operation in science and engineering. Calculating area and volume, velocity from acceleration, work from force and displacement are just a few examples where integrals are used. Integration of simple functions can be done analytically, but more involved functions are frequently difficult or impossible to integrate analytically. In calculus courses the integrand (the quantity to be integrated) is usually a function. In applications of science and engineering the integrand can be a function or a set of data points. For example, data points from discrete measurements of flow velocity can be used to calculate volume.

It is assumed in the presentation below that the reader has knowledge of integrals and integration. A definite integral of a function $f(x)$ from a to b has the form:

$$q = \int_a^b f(x)dx$$

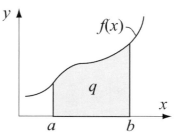

The function $f(x)$ is called the integrand and the numbers a and b are the limits of integration. Graphically, the value of the integral q is the area between the graph of the function, the x axis, and the limits a and b (the shaded area in the figure). When a definite integral is calculated analytically $f(x)$ is always a function. When the integral is calculated numerically $f(x)$ can be a function or a set of points. In numerical integration the total area is obtained by dividing the area into small sections, calculating the area of each section, and adding them up. Various numerical methods have been developed for this purpose. The difference between the methods is in the way that the area is divided into sections and the method by which the area of each section is calculated. The reader is referred to books on numerical analysis for details of the numerical techniques.

The following describes how to use the three MATLAB built-in integration functions `quad`, `quadl` and `trapz`. The `quad` and `quadl` commands are used for integration when $f(x)$ is a function, while the `trapz` is used when $f(x)$ is given by data points.

The `quad` command:

The form of the `quad` command which uses the adaptive Simpson method of integration is:

The value of the integral.   The function to be integrated.   The integration limits.

- The function can be entered as a string expression, or as a function handle, in the same way as with the `fzero` command. See Section 10.1 for details. The first two methods are demonstrated in Sample Problem 10-2 on the following page.

- The function $f(x)$ must be written for an argument x that is a vector (use element-by-element operations), such that it calculates the value of the function for each element of x.

- The user has to make sure that the function does not have a vertical asymptote between a and b.

- `quad` calculates the integral with an absolute error that is smaller than 1.0e–6. This number can be changed by adding an optional `tol` argument to the command:

      q = quad('function',a,b,tol)

  `tol` is a number that defines the maximum error. With larger `tol` the integral is calculated less accurately, but faster.

The `quadl` command:

The form of the `quadl` (the last letter is a lower case L) command is exactly the same as the `quad` command:

The value of the integral.   The function to   The integration limits.
                                 be integrated.

All the comments that are listed above for the `quad` command are valid for the `quadl` command. The difference between the two commands is in the numerical method that is used for calculating the integration. The `quadl` uses the adaptive Lobatto method, which can be more efficient for high accuracies and smooth integrals.

Sample Problem 10-2: Numerical integration of a function

Use numerical integration to calculate the following integral:

$$\int_0^8 (xe^{-x^{0.8}} + 0.2)\,dx$$

Solution

For illustration, a plot of the function for the interval $0 \leq x \leq 8$ is shown on the right. The solution uses the quad command and shows how to enter the function in the command in two ways. In the first, it is entered directly by typing the expression as an argument. In the second, an anonymous function is created first and its name is subsequently entered in the command.

The use of the quad command in the Command Window with the function to be integrated typed in as a string, is shown below. Note that the function is typed with element-by-element operations.

```
>> quad('x.*exp(-x.^0.8)+0.2',0,8)

ans =
    3.1604
```

The second method is to first create a user-defined function that calculates the function to be integrated. The function file (named y=Chap10Sam2(x)) is:

```
function y=Chap10Sam2(x)
y=x.*exp(-x.^0.8)+0.2;
```

Note again that the function is written with element-by-element operations such that the argument x can be a vector. The integration is then done in the Command Window by typing the handle @Chap10Sam2 for the argument function in the quad as shown below:

```
>> q=quad(@Chap10Sam2,0,8)

q =
    3.1604
```

The trapz command:

The trapz command can be used for integrating a function that is given as data points. It uses the numerical trapezoidal method of integration. The form of the command is:

$$q = \text{trapz}(x,y)$$

where x and y are vectors with the x and y coordinates of the points, respectively. The two vectors must be of the same length.

10.4 ORDINARY DIFFERENTIAL EQUATIONS

Differential equations play a crucial role in science and engineering since they are in the foundation of virtually every physical phenomena that is involved in engineering applications. Only a limited number of differential equations can be solved analytically. Numerical methods, on the other hand, can give an approximate solution to almost any equation. Obtaining a numerical solution might not be, however, a simple task. This is because a numerical method that can solve any equation does not exist. Instead, there are many methods that are suitable for solving different types of equations. MATLAB has a large library of tools that can be used for solving differential equations. To fully utilize the power of MATLAB, however, requires that the user have knowledge of differential equations and the various numerical methods that can be used for solving them.

This section describes in detail how to use MATLAB to solve a first order ordinary differential equation. The possible numerical methods that can be used for solving such an equation are mentioned and described in general terms, but are not explained from a mathematical point of view. This section provides information for solving simple, "nonproblematic," first-order equations. This solution provides the basis for solving higher-order equations and a system of equations.

An ordinary differential equation (ODE) is an equation that contains an independent variable, a dependent variable, and derivatives of the dependent variable. The equations that are considered here are of first order with the form:

$$\frac{dy}{dx} = f(x, y)$$

where x and y are the independent and dependent variables, respectively. A solution is a function $y = f(x)$ that satisfies the equation. In general, many functions can satisfy a given ODE and more information is required for determining the solution of a specific problem. The additional information is the value of the function (the dependent variable) at some value of the independent variable.

Steps for solving a single first order ODE:

For the remainder of this section the independent variable is taken as t (time). This is done because in many applications time is the independent variable, and to be consistent with the information in the **Help** menu of MATLAB.

Step 1: **Write the problem in a standard form.**

Write the equation in the form:

$$\frac{dy}{dt} = f(t, y) \quad \text{for } t_0 \le t \le t_f, \text{ with } y = y_0 \text{ at } t = t_0.$$

As shown above, three pieces of information are needed for solving a first order ODE: An equation that gives an expression for the derivative of y with respect to t, the interval of the independent variable, and the initial value of y. The solution is the value of y as a function of t between t_0 and t_f.

An example of a problem to solve is:

$$\frac{dy}{dt} = \frac{t^3 - 2y}{t} \quad \text{for} \quad 1 \le t \le 3 \quad \text{with} \quad y = 4.2 \quad \text{at} \quad t = 1.$$

Step 2:   Create a user defined function (in a function file), or an anonymous function.

The ODE to be solved has to written as a user-defined function (in a function file), or as an anonymous function. Both calculate $\frac{dy}{dt}$ for given values of t and y. For the example problem above, the user-defined function (which is saved as a separate file) is:

```
function dydt=ODEexp1(t,y)
dydt=(t^3-2*y)/t;
```

When an anonymous function is used, it can be defined in the Command Window, or be within a script file. For the example problem here the anonymous function (names ode) is:

```
>> ode1=@(t,y)(t^3-2*y)/t
ode1 =
    @(t,y)(t^3-2*y)/t
```

Step 3:   Select a method of solution.

Select the numerical method that you would like MATLAB to use in the solution. Many numerical methods have been developed to solve first order ODE's, and several of the methods are available as built-in functions in MATLAB. In a typical numerical method, the time interval is divided into small time steps. The solution starts at the known point y_0, and then by using one of the integration methods the value of y is calculated at each time step. Table 10-1 lists seven ODE solver commands, which are MATLAB built-in functions that can be used for solving a first order ODE. A short description of each solver is included in the table.

Table 10-1: MATLAB ODE Solvers

ODE Solver Name	Description
ode45	For nonstiff problems, one step solver, best to apply as a first try for most problems. Based on explicit Runge-Kutta method.
ode23	For nonstiff problems, one step solver. Based on explicit Runge-Kutta method. Often quicker but less accurate than ode45.
ode113	For nonstiff problems, multistep solver.

Table 10-1: MATLAB ODE Solvers

ODE Solver Name	Description
ode15s	For stiff problems, multistep solver. Use if ode45 failed. Uses a variable order method.
ode23s	For stiff problems, one step solver. Can solve some problems that ode15 cannot.
ode23t	For moderately stiff problems.
ode23tb	For stiff problems. Often more efficient than ode15s.

In general, the solvers can be divided into two groups according to their ability to solve stiff problems and according to whether they use one or multistep methods. Stiff problems are ones that include fast and slowly changing components and require small time steps in their solution. One-step solvers use information from one point to obtain solution at the next point. Multistep solvers use information from several previous points to find the solution at the next point. The details of the different methods are beyond the scope of this book.

It is impossible to know ahead of time which solver is the most appropriate for a specific problem. A suggestion is to first try ode45 which gives good results for many problems. If a solution is not obtained because the problem is stiff, it is suggested to try the solver ode15s.

Step 4:   Solve the ODE.

The form of the command that is used to solve an initial value ODE problem is the same for all the solvers and all the equations that are solved. The form is:

$$[t, y] = \texttt{solver\_name(ODEfun, tspan, y0)}$$

Additional information:

solver_name   Is the name of the solver (numerical method) that is used (e.g. ode45 or ode23s)

ODEfun   The function from **Step 2** that calculates $\frac{dy}{dt}$ for given values of t and y. If it was written as a user-defined function, the function handle is entered. If it was written as an anonymous function, the name of the anonymous function is entered. (See example that follows.)

tspan   A vector that specifies the interval of the solution. The vector must have at least two elements, but can have more. If the vector has only two elements, the elements must be [t0 tf], which are the initial and final points of the solution interval.

The vector `tspan` can have, however, additional points between the first and last points. The number of elements in `tspan` affects the output from the command. See `[t,y]` below.

y0   Is the initial value of *y* (the value of *y* at the first point of the interval.)

[t,y]   Is the output, which is the solution of the ODE. `t` and `y` are column vectors. The first and the last points are the beginning and end points of the interval. The spacing and number of points in between depends on the input vector `tspan`. If `tspan` has two elements (the beginning and end points) the vectors `t` and `y` contain the solution at every integration step calculated by the solver. If `tspan` has more than two points (additional points between the first and the last) the vectors `t` and `y` contain the solution only at these points. The number of points in `tspan` does not affect the time steps used for the solution by the program.

For example, consider the solution to the problem stated in **Step 1**:

$$\frac{dy}{dt} = \frac{t^3 - 2y}{t} \quad \text{for } 1 \le t \le 3 \quad \text{with } y = 4.2 \text{ at } t = 1,$$

If the ode function is written as a user-defined function (see **Step 2**), then the solution with MATLABs built-in function `ode45` is obtained by:

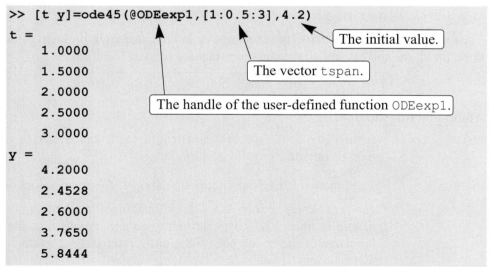

The solution is obtained with the solver `ode45`. The name of the user-defined function from step 2 is `ODEexp1`. The solution starts at *t* = 1 and ends at *t* = 3 with increments of 0.5 (according to the vector `tspan`). To show the solution, the problem is solved again below using `tspan` with smaller spacing, and the solution is plotted with the `plot` command.

```
>> [t y]=ode45(@ODEexp1,[1:0.01:3],4.2);
>> plot(t,y)
>> xlabel('t'), ylabel('y')
```

The plot that is created is:

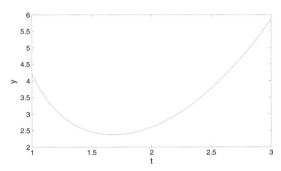

If the ode function is written as an anonymous function called `ode1` (see **Step 2**), then the solution (same as shown above) is obtained by typing:
`[t y]=ode45(ode1,[1:0.5:3],4.2)`.

10.5 EXAMPLES OF MATLAB APPLICATIONS

Sample Problem 10-3: The gas equation

The ideal gas equation relates the volume (V in L), temperature (T in K), pressure (P in atm), and the amount of gas (number of moles n) by:

$$p = \frac{nRT}{V}$$

where $R = 0.08206$ L-atm/mol-K is the gas constant.

The van der Waals equation gives the relationship between these quantities for a real gas by:

$$\left(P + \frac{n^2 a}{V^2}\right)(V - nb) = nRT$$

where a and b are constants that are specific for each gas.

Use the `fzero` function to calculate the volume of 2 mol CO_2 at temperature of 50°C, and pressure of 6 atm. For CO_2, $a = 3.59$ L^2-atm/mol^2, and $b = 0.0427$ L/mol.

Solution

The solution written in a script file is shown below.
The program first calculates an estimated value of the volume using the ideal gas

```
global P T n a b R
R=0.08206;
P=6; T=323.2; n=2; a=3.59; b=0.047;
Vest=n*R*T/P;
V=fzero(@Waals,Vest)
```

Calculating an estimated value for *V*.

Function handle @waals is used to pass the user-defined function waals into fzero.

equation. This value is then used in the fzero command for the estimate of the solution. The van der Waals equation is written as a user-defined function named Waals, which is shown below:

```
function fofx=Waals(x)
global P T n a b R
fofx=(P+n^2*a/x^2)*(x-n*b)-n*R*T;
```

In order for the script and function files to work correctly, the variables P, T, n, a, b, and R are declared global. When the script file (saved as Chap10SamPro3) is executed in the Command Window the value of *V* is displayed, as shown next:

```
>> Chap10SamPro3
V =
    8.6613
```

The volume of the gas is 8.6613 L.

Sample Problem 10-4: Maximum viewing angle

To get the best view of a movie, a person has to sit at a distance *x* from the screen such that the viewing angle θ is maximum. Determine the distance *x* for which θ is maximum for the configuration shown in the figure.

Solution

The problem is solved by writing a function for the angle θ in terms of *x*, and then finding the *x* for which the angle is maximum. In the triangle that includes θ, one side is given (the height of the screen), and the other two sides can be written in terms of *x*, as shown in the figure. One way in which θ can be written in terms of *x* is by using the law of cosines.

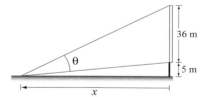

$$\cos(\theta) = \frac{(x^2 + 5^2) + (x^2 + 41^2) - 36^2}{2\sqrt{x^2 + 5^2}\sqrt{x^2 + 41^2}}$$

The angle θ is expected to be between 0 and $\pi/2$. Since $\cos(0) = 1$ and the cosine is decreasing with increasing θ, the maximum angle corresponds to the smallest $\cos(\theta)$. A plot of $\cos(\theta)$ as a function of x shows that the function has a minimum between 10 and 20. The commands for the plot are:

```
>>fplot('((x^2+5^2)+(x^2+41^2)-36^2)/(2*sqrt(x^2+ 5^2)*sqrt(x^2+
                                                  41^2))',[0 25])

>> xlabel('x'); ylabel('cos(\theta)')
```

The minimum can be determined with the `fminbnd` command:

```
>>[x anglecos]=fminbnd('((x^2+5^2)+(x^2+41^2)-36^2)/
                   (2*sqrt(x^2+5^2)*sqrt(x^2+41^2))',10,20)
x =
    14.3178
anglecos =
    0.6225
```

The minimum is at $x = 14.3178$ m.
At this point $\cos(\theta) = 0.6225$.

```
>> angle=anglecos*180/pi
angle =
    35.6674
```

In degrees the angle is $35.6674°$.

Sample Problem 10-5: Water flow in a river

To estimate the amount of water that flows in a river during a year, a section of the river is made up to have a rectangular cross section as shown. In the beginning of every month (starting at January 1st) the height h of the water, and the speed v of the water flow is measured. The first day of measurement is taken as 1, and the last day which is January 1st of the next year is day 366. The following data was measured:

Day	1	32	60	91	121	152	182	213	244	274	305	335	366
h (m)	2.0	2.1	2.3	2.4	3.0	2.9	2.7	2.6	2.5	2.3	2.2	2.1	2.0
v (m/s)	2.0	2.2	2.5	2.7	5	4.7	4.1	3.8	3.7	2.8	2.5	2.3	2.0

Use the data to calculate the flow rate, and then integrate the flow rate to obtain an estimate of the total amount of water that flows in the river during a year.

Solution

The flow rate, Q, (volume of water per second) at each data point is obtained by multiplying the water speed by the width and height of the cross-sectional area of the water that flows in the channel:

$$Q = vwh \quad (m^3/s)$$

The total amount of water that flows is estimated by the integral:

$$V = (60 \cdot 60 \cdot 24) \int_{t_1}^{t_2} Q dt$$

The flow rate is given in cubic meters per second which means that the time must have units of seconds. Since the data is given in terms of days, the integral is multiplied by $(60 \cdot 60 \cdot 24)$ s/day.

The following is a program written in a script file that first calculates Q and then carries out the integration using the `trapz` command. The program also generates a plot of the flow rate vs. time.

```
w=8;
d=[1 32 60 91 121 152 182 213 244 274 305 335 366];
h=[2 2.1 2.3 2.4 3.0 2.9 2.7 2.6 2.5 2.3 2.2 2.1 2.0];
speed=[2 2.2 2.5 2.7 5 4.7 4.1 3.8 3.7 2.8 2.5 2.3 2];
Q=speed.*w.*h;
Vol=60*60*24*trapz(d,Q);
fprintf('The estimated amount of water that flows in the
river in a year is %g cubic meters.',Vol)
plot(d,Q)
xlabel('Day'), ylabel('Flow Rate (m^3/s)')
```

When the file (saved as Chap10SamPro5) is executed in the Command Window, the estimated amount of water is displayed and the plot is generated. Both are shown below:

```
>> Chap10SamPro5
The estimated amount of water that flows in the river in a
year is 2.03095e+009 cubic meters.
```

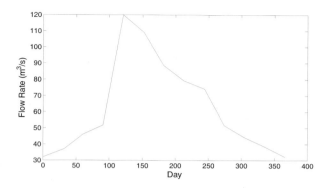

Sample Problem 10-6: Car crash into a safety bumper

A safety bumper is placed at the end of a racetrack to stop out-of-control cars. The bumper is designed such that the force that the bumper applies to the car is a function of the velocity v and the displacement x of the front edge of the bumper according to the equation:

$$F = Kv^3(x+1)^3$$

where $K = 30$ s-kg/m^5 is a constant.

A car with a mass m of 1500 kg hits the bumper at a speed of 90 km/h. Determine and plot the velocity of the car as a function of its position for $0 \le x \le 3$ m.

Solution

The deceleration of the car once it hits the bumper can be calculated from Newton's second law of motion.

$$ma = -Kv^3(x+1)^3$$

which can be solved for the acceleration a as a function of v and x:

$$a = \frac{-Kv^3(x+1)^3}{m}$$

The velocity as a function of x can be calculated by substituting the acceleration in the equation:

$$vdv = adx$$

which gives:

$$\frac{dv}{dx} = \frac{-Kv^2(x+1)^3}{m}$$

The last equation is a first order ODE that needs to be solved for the interval $0 \le x \le 3$ with the initial condition: $v = 90$ km/h at $x = 0$.

A numerical solution of the differential equation with MATLAB is shown in the following program which is written in a script file:

```
global k m
k=30; m=1500; v0=90;
xspan=[0:0.2:3];          A vector that specifies the interval of the solution.
vOmps=v0*1000/3600;            Changing the units of v0 to m/s.
[x v]=ode45(@bumper,xspan,vOmps)        Solving the ODE.
plot(x,v)
xlabel('x (m)'); ylabel('velocity (m/s)')
```

Note that a function handle @bumper is used for passing the user-defined function bumper into ode45. The listing of the user-defined function with the differential equation, named bumper, is:

```
function dvdx=bumper(x,v)
global k m
dvdx=-(k*v^2*(x+1)^3)/m;
```

When the script file executes (was saved as Chap10SamPro6) the vectors x and v
are displayed in the Command Window (actually they are displayed on the screen
one after the other, but to save room they are displayed below next to each other).

```
>> Chap10SamPro6
x =              v =
          0          25.0000
     0.2000          22.0420
     0.4000          18.4478
     0.6000          14.7561
     0.8000          11.4302
     1.0000           8.6954
     1.2000           6.5733
     1.4000           4.9793
     1.6000           3.7960
     1.8000           2.9220
     2.0000           2.2737
     2.2000           1.7886
     2.4000           1.4226
     2.6000           1.1435
     2.8000           0.9283
     3.0000           0.7607
```

The plot of the velocity as a function of distance generated by the program is:

10.6 PROBLEMS

1. Determine the solution of the equation: $10(e^{-0.2x} - e^{-1.5x}) = 4$.

2. Determine the three positive roots of the equation: $e^{-0.2x}\cos(2x) = 0.15x^2 - 1$.

3. Determine the positive roots of the equation: $x^3 - 7x^2\cos(3x) + 3 = 0$.

4. A box of mass $m = 18\,\text{kg}$ is being pulled by a rope as shown. The force that is required to move the box is given by:

$$F = \frac{\mu mg\sqrt{x^2 + h^2}}{x + \mu h}$$

where $h = 10\,\text{m}$, $\mu = 0.55$ is the friction coefficient, and $g = 9.81\,\text{m/s}^2$. Determine the distance x when the pulling force is equal to 90 N.

5. A scale is made of three springs, as shown in the figure. Initially, the springs are not stretched. When an object is attached to the ring, the springs stretch and the ring is displaced down a distance x. The weight of the object can be expressed in terms of the distance x by:

$$W = \frac{2K}{L}(L - L_0)(b + x) + \frac{K}{2}x$$

where $L_0 = \sqrt{a^2 + b^2}$ is the initial length of a diagonal spring, and

$L = \sqrt{a^2 + (b + x)^2}$ is the stretched length of a diagonal spring.

For the given scale $a = 0.18\,\text{m}$, $b = 0.06\,\text{m}$, and the springs' constant is $K = 2600\,\text{N/m}$. Determine the distance x when a 200 N object is attached to the scale.

6. The diode in the circuit shown is forward biased. The current I flowing through the diode is given by:

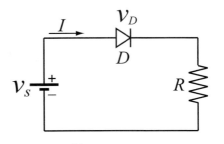

$$I = I_S \left(e^{\frac{q v_D}{kT}} - 1 \right)$$

where v_D is the voltage drop across the diode, T is the temperature in Kelvins, $I_S = 10^{-12}$ A is the saturation current, $q = 1.6 \times 10^{-16}$ coulombs is the elementary charge value, and $k = 1.38 \times 10^{-23}$ joule/K is Boltzmann's constant. The current I flowing through the circuit (the same as the current in the diode) is given also by:

$$I = \frac{v_S - v_D}{R}$$

Determine v_D if $v_S = 2$ V, $T = 297$ K, and $R = 1000\,\Omega$. (Substitute I from one equation into the other equation and solve the resulting nonlinear equation.)

7. Determine the minimum of the function: $f(x) = 10(e^{-0.2x} - e^{-1.5x}) - 4$.

8. A paper cup shaped as a frustum of a cone with $R_2 = 1.3R_1$ is designed to have a volume of 240 cm^3. Determine R_1 and height of the cup such that the least amount of paper will be used for making the cup.

9. Consider again the box that is being pulled in Problem 4. Determine the distance x at which the force that is required to pull the box is the smallest. What is the magnitude of this force?

10. Determine the dimensions (radius r and height h) and the volume of the cone with the largest volume that can be made inside of a sphere with a radius R of 17 cm.

11. Planck's radiation law gives the spectral radiancy R as a function of the wave length λ and temperature T (in degrees K):

$$R = \frac{2\pi c^2 h}{\lambda^5} \frac{1}{e^{(hc)/(\lambda kT)} - 1}$$

where $c = 3.0 \times 10^8$ m/s is the speed of light, $h = 6.63 \times 10^{-34}$ J-s is the Planck constant, and $k = 1.38 \times 10^{-23}$ J/K is the Boltzmann constant.

Plot R as a function of λ for $0.2 \times 10^{-6} \le \lambda \le 6.0 \times 10^{-6}$ m at $T = 1500$ K, and determine the wavelength that gives the maximum R at this temperature.

12. Use MATLAB to calculate the following integral:

$$\int_1^6 \frac{3 + e^{0.5x}}{0.3x^2 + 2.5x + 1.6} dx$$

13. The length L of the main supporting cable of a suspension bridge can be calculated by:

$$L = 2\int_0^a \left(1 + \frac{4h^2}{a^4}x^2\right)^{1/2} dx$$

where a is half the length of the bridge, and h is the distance from the deck to the top of the tower where the cable is attached. Determine the length of a bridge with $a = 60$ m, and $b = 15$ m.

14. The electric field E due to a charged circular disk at a point at a distance z along the axis of the disk is given by:

$$E = \frac{\sigma z}{4\varepsilon_0} \int_0^R (z^2 + r^2)^{-3/2}(2r)dr$$

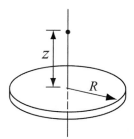

where σ is the charge density, ε_0 is the permittivity constant, $\varepsilon_0 = 8.85 \times 10^{-12}$ C^2/N-m^2, and R is the radius of the disk. Determine the electric field at a point located 5 cm from a disk with a radius of 6 cm, charged with $\sigma = 300$ μC/m^2.

15. The variation of gravitational acceleration g with altitude y is given by:

$$g = \frac{R^2}{(R + y)^2}g_0$$

where $R = 6371$ km is the radius of the earth, and $g_0 = 9.81$ m/s^2 is the gravitational acceleration at sea level. The change in the gravitational potential energy, ΔU, of an object that is raised up from the Earth is given by:

$$\Delta U = \int_0^h mg\,dy$$

Determine the change in the potential energy of a satellite with a mass of 500 kg that is raised from the surface of the earth to a height of 800 km.

16. An approximate map of the state of Illinois is shown in the figure. Measurements of the width of the state are marked at intervals of 30 miles. Use numerical integration to estimate the area of the state. Compare the result with the actual area of Illinois, which is 57,918 square miles.

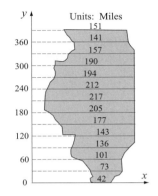

17. The time-dependent relaxation modulus $G(t)$ of many biological materials can be described by Fung's reduced relaxation function:

$$G(t) = G_\infty \left(1 + c \int_{\tau_1}^{\tau_2} \frac{e^{(-t)/x}}{x}\,dx \right)$$

Use numerical integration to find the relaxation modulus at 10 s, 100 s, and 1000 s. Assume $G_\infty = 5\,\text{ksi}$, $c = 0.05$, $\tau_1 = 0.05\,\text{s}$, and $\tau_2 = 500\,\text{s}$.

18. The orbit of Pluto is elliptical in shape, with $a = 5.9065 \times 10^9\,\text{km}$ and $b = 5.7208 \times 10^9\,\text{km}$. The perimeter of an ellipse can be calculated by:

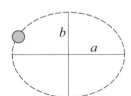

$$P = 4a \int_0^{\pi/2} \sqrt{1 - k^2 \sin^2\theta}\,d\theta$$

where $k = \dfrac{\sqrt{a^2 - b^2}}{a}$. Determine the distance Pluto travels in one orbit. Calculate the average speed that Pluto travels (in km/h) if one orbit takes about 248 years.

19. Solve:

$$\frac{dy}{dx} = x + \frac{xy}{3} \quad \text{for } 1 \le x \le 4 \text{ with } y(1) = 2.$$

Plot the solution.

20. Solve:

$$\frac{dy}{dx} = 0.6x\sqrt{y} + 0.5y\sqrt{x} \quad \text{for } 0 \le x \le 4 \text{ with } y(0) = 2.5.$$

Plot the solution.

21. A water tank shaped as an inverted frustum cone has a circular hole at the bottom on the side, as shown. According to Torricelli's law, the speed v of the water that is discharging from the hole is given by:

$$v = \sqrt{2gh}$$

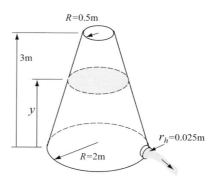

where h is the height of the water and $g = 9.81 \text{ m/s}^2$. The rate at which the height, y, of the water in the tank changes as the water flows out through the hole is given by:

$$\frac{dy}{dt} = \frac{\sqrt{2gy}r_h^2}{(2 - 0.5y)^2}$$

where r_h is the radius of the hole.

Solve the differential equation for y. The initial height of the water is $y = 2 \text{ m}$. Solve the problem for different times and find the time where $y = 0.1 \text{ m}$. Make a plot of y as a function of time.

22. The sudden outbreak of an insect population can be modeled by the equation:

$$\frac{dN}{dt} = RN\left(1 - \frac{N}{C}\right) - \frac{rN^2}{N_c^2 + N^2}$$

The first term relates to the well known logistic population growth model where N is the number of insects, R is an intrinsic growth rate, and C is the carrying capacity of the local environment. The second term represents the effects of bird predation. Its effect becomes significant when the population reaches a critical size N_c. r is the maximum value that the second term can reach at large values of N.

Solve the differential equation for $0 \le t \le 50$ days and two growth rates, $R = 0.55$ and $R = 0.58$ days^{-1}, and with $N(0) = 10,000$. The other parameters are $C = 10^4$, $N_c = 10^4$, $r = 10^4$ 1/day. Make one plot comparing the two solutions and discuss why this model is called an "outbreak" model.

23. The growth of a fish is often modeled by the von Bertalanffy growth model:

$$\frac{dw}{dt} = aw^{2/3} - bw$$

where w is the weight and a and b are constants. Solve the equation for w for the case $a = 5 \text{ lb}^{1/3}$, $b = 2 \text{ day}^{-1}$, and $w(0) = 0.5 \text{ lb}$. Make sure that the selected time span is just long enough so that the maximum weight is approached. What is the maximum weight for this case? Make a plot of w as a function of time.

24. An airplane uses a parachute and other means of braking as it slows down on the runway after landing. Its acceleration is given by $a = -0.0035v^2 - 3$ m/s². Since $a = \dfrac{dv}{dt}$, the rate of change of the velocity is given by:

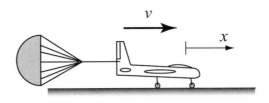

$$\frac{dv}{dt} = -0.0035v^2 - 3$$

Consider an airplane with a velocity of 300 km/h that opens its parachute and starts decelerating at $t = 0$ s.

a) By solving the differential equation, determine and plot the velocity as a function of time from $t = 0$ s until the airplane stops.

b) Use numerical integration to determine the distance x the airplane travels as a function of time. Make a plot of x vs. time.

25. An RC circuit includes a voltage source v_s, a resistor $R = 50$ Ω, and a capacitor $C = 0.001$ F, as shown in the figure. The differential equation that describes the response of the circuit is:

$$\frac{dv_c}{dt} + \frac{1}{RC}v_c = \frac{1}{RC}v_s$$

where v_c is the voltage of the capacitor. Initially, $v_s = 0$, and then at $t = 0$ the voltage source is changed. Determine the response of the circuit for the following three cases:

a) $v_s = 12$ V for $t \geq 0$.

b) $v_s = 12\sin(2 \cdot 60\pi t)$ V for $t \geq 0$.

c) $v_s = 12$ V for $0 \leq t \leq 0.01$ s, and then $v_s = 0$ for $t \geq 0.01$ s (rectangular pulse).

Each case corresponds to a different differential equation. The solution is the voltage of the capacitor as a function of time. Solve each case for $0 \leq t \leq 0.2$ s. For each case plot v_s and v_c vs. time (make two separate plots on the same page).

Chapter 11
Symbolic Math

All the mathematical operations done with MATLAB in the first ten chapters were numerical. The operations were carried out by writing numerical expressions that could contain numbers and variables with preassigned numerical values. When a numerical expression is executed by MATLAB, the outcome is also numerical (a single number or an array with numbers). The number, or numbers, are either exact or a floating point-approximated value. For example, typing 1/4 gives 0.2500 – an exact value, and typing 1/3 gives 0.3333 – an approximated value.

Many applications in math, science, and engineering require symbolic operations, which are mathematical operations with expressions that contain symbolic variables (variables that don't have specific numerical values when the operation is executed). The result of such operations is also a mathematical expression in terms of the symbolic variables. One simple example is solving an algebraic equation which contains several variables for one variable in terms of the others. If a, b, and x are symbolic variables, and $ax - b = 0$, x can be solved in terms of a and b to give $x = b/a$. Other examples of symbolic operations are analytical differentiation or integration of mathematical expressions. For instance, the derivative of $2t^3 + 5t - 8$ with respect to t is $6t^2 + 5$.

MATLAB has the capability of carrying out many types of symbolic operations. The numerical part of the symbolic operation is carried out by MATLAB exactly without approximating numerical values. For example, the result of adding $\frac{x}{4}$ and $\frac{x}{3}$ is $\frac{7}{12}x$ and not $0.5833x$.

Symbolic operations can be performed by MATLAB when the Symbolic Math Toolbox is installed. The Symbolic Math Toolbox is a collection of MATLAB functions that are used for execution of symbolic operations. The commands and functions for the symbolic operations have the same style and syntax as those for the numerical operations. The symbolic operations themselves are executed primarily by Maple®, which is mathematical software designed for this purpose. The Maple software is embedded within MATLAB and is automatically activated when a symbolic MATLAB function is executed. Maple also exists as separate independent software. That software, however, has a completely different structure and commands than MATLAB. The Symbolic Math Toolbox is included in the student version of MATLAB. In the standard version, the toolbox is purchased separately. To check if the Symbolic Math Toolbox is installed on a computer, the user can type the command `ver` in the Command Window. In response, MAT-LAB displays information about the version that is used as well as a list of the tool boxes that are installed.

The starting point for symbolic operations is symbolic objects. Symbolic objects are made of variables and numbers that, when used in mathematical expressions, tell MATLAB to execute the expression symbolically. Typically, the user first defines (creates) the symbolic variables (objects) that are needed, and then uses them to create symbolic expressions which are subsequently used in symbolic operation. If needed, symbolic expressions can be used in numerical operations

The first section in this chapter describes how to define symbolic objects and how to use them to create symbolic expressions. The second section shows how to change the form of existing expressions. Once a symbolic expression exists, it can be used in mathematical operations. MATLAB has a large selection of functions for this purpose. The next four sections (11.3–11.6) describe how to use MATLAB to solve algebraic equations, to carry out differentiation and integration, and to solve differential equations. Section 11.7 covers plotting symbolic expressions. How to use symbolic expressions in subsequent numerical calculations is explained in the following section.

11.1 SYMBOLIC OBJECTS, AND SYMBOLIC EXPRESSIONS

A symbolic object can be a variable (without a preassigned numerical value), a number, or an expression made of symbolic variables and numbers. Symbolic expression is a mathematical expression containing one or more symbolic objects. When typed, a symbolic expression may look like a standard numerical expression. However, because the expression contains symbolic objects, it is executed by MATLAB symbolically.

11.1.1 Creating Symbolic Objects

Symbolic objects can be variables or numbers. They can be created with the `sym` and/or the `syms` commands. A single symbolic object can be created with the `sym` command:

```
object_name = sym('string')
```

where the string which is the symbolic object is assigned to a name. The string can be:

- A single letter, or a combination of several letters (no spaces). Examples: `'a'`, `'x'`, or `'yad'`.

- A combination of letters and digits starting with a letter and with no spaces Examples: 'xh12', or 'r2d2'.

- A number. Examples: '15', or '4'.

In the first two cases (when the string is a single letter, a combination of several letters, or a combination of letters and digits), the symbolic object is a symbolic variable. In this case it is convenient (but not necessary) to name the object the same as the string. For example, *a*, *bb*, and *x*, can be defined as symbolic variables by:

```
>> a=sym('a')          Create a symbolic object a and assign it to a.

a =

a
>> bb=sym('bb')        The display of a symbolic
                        object is not indented.
bb =
bb
>> x=sym('x');         The symbolic variable x is created but not displayed,
>>                      since a semicolon is typed at the end of the command.
```

The name of the symbolic object can be different than the name of the variable. For example:

```
>> g=sym('gamma')      The symbolic object is gamma, and
                        the name of the object is g.
g =
gamma
```

As mentioned, symbolic objects can also be numbers. The numbers don't have to be typed as strings. For example, the sym command is used next to create symbolic objects from the numbers 5 and 7 and assign them to the variables *c* and *d*, respectively.

```
>> c=sym(5)      Create a symbolic object from the number 5 and assign it to c.

c =
5
                  The display of a symbolic
>> d=sym(7)       object is not indented.

d =
7
```

As shown, when a symbolic object is created and a semicolon is not typed at the end of the command, MATLAB displays the name of the object and the object itself in the next two lines. The display of symbolic objects starts at the beginning of the line and is not indented like the display of numerical variables. The difference is illustrated below where a numerical variable is created.

```
>> e=13
```
13 is assigned to e (numerical variable).

```
e =
    13
```
The display of the value of a numerical variable is indented.

Several symbolic variables can be created in one command by using the `syms` command which has the form:

```
syms variable_name variable_name variable_name
```

The command creates symbolic objects that have the same name as the symbolic variables. For example, the variables y, z, and d can all be created as symbolic variables in one command by typing:

```
>> syms y z d
>> y
y =
y
```
The variables created by the `syms` command are not displayed automatically. Typing the name of the variable shows that the variable was created,

When the `syms` command is executed, the variables it creates are not displayed automatically – even if a semicolon is not typed at the end of the command.

11.1.2 Creating Symbolic Expressions

Symbolic expressions are mathematical expressions written in terms of symbolic variables. Once symbolic variables are created, they can be used for creating symbolic expressions. The symbolic expression is a symbolic object (the display is not indented). The form for creating a symbolic expression is:

```
Expression_name = Mathematical expression
```

A few examples are:

```
>> syms a b c x y
```
Define a, b, c, x, and y as symbolic variables.

```
>> f=a*x^2+b*x + c

f =
a*x^2+b*x+c
```
Create the symbolic expression $ax^2 + bx + c$ and assign it to f.

The display of the symbolic expression is not indented.

When a symbolic expression, which includes mathematical operations that can be executed (addition, subtraction, multiplication, and division), is entered, MAT-LAB executes the operations as the expression is created. For example:

```
>> g=2*a/3+4*a/7-6.5*x+x/3+4*5/3-1.5
```
$\dfrac{2a}{3} + \dfrac{4a}{7} - 6.5x + \dfrac{x}{3} + 4 \cdot \dfrac{5}{3} - 1.5$ is entered.

```
g =
26/21*a-37/6*x+31/6
```
$\dfrac{26}{21}a - \dfrac{37}{6}x + \dfrac{31}{6}$ is displayed.

Notice that all the calculations are carried out exactly with no numerical approximation. In the last example, $\dfrac{2}{3}a$ and $\dfrac{4}{7}a$ were added by MATLAB to give $\dfrac{26}{21}a$, and $-6.5x + \dfrac{x}{3}$ was added to $\dfrac{37}{6}x$. The operations with the terms that contain only numbers in the symbolic expression are carried out exactly. In the last example, $4 \cdot \dfrac{5}{3} + 1.5$ is replaced by $\dfrac{31}{6}$.

The difference between exact and approximate calculations is demonstrated in the following example, where the same mathematical operations are carried out once with symbolic variables and once with numerical variables.

```
>> a=sym(3); b=sym(5);
```
Define a and b as symbolic 3 and 5, respectively.

```
>> e=b/a+sqrt(2)
```
Create an expression that includes a and b.

```
e =
5/3+2^(1/2)
```
An exact value of e is displayed as a symbolic object (The display is not indented).

```
>> c=3; d=5;
```
Define c and d as numerical 3 and 5, respectively.

```
>> f=d/c+sqrt(2)
```
Create an expression that includes c and d.

```
f =
    3.0809
```
An approximated value of f is displayed as a number (The display is indented).

An expression that is created can include both symbolic objects and numerical variables. However, if an expression includes a symbolic object (one or several) all the mathematical operations will be carried out exactly. For example, if c is replaced by a in the last expression, the result is exact as it was in the first example.

```
>> g=d/a+sqrt(2)

g =
5/3+2^(1/2)
```

Additional facts about symbolic expressions and symbolic objects:

- Symbolic expressions can include numerical variables which have been obtained from the execution of numerical expressions. When these variables are inserted in symbolic expressions their exact value is used, even if the variable was displayed before with an approximated value. For example:

```
>> h=10/3
```
h is defined to be 10/3 (a numerical variable).

```
h =
    3.3333
```
An approximated value of h (numerical variable) is displayed.

```
>> k=sym(5); m=sym(7);        Define k and m as symbolic 5 and 7, respectively.

>> p=k/m+h                                  h, k, and m are used in an expression.

p =                           The exact value of h is used in the determination of p.
85/21                         An exact value of p (symbolic object) is displayed.
```

- The `double(S)` command can be used to convert a symbolic expression (object) S which is written in an exact form to numerical form. (The name double comes from the fact that the command returns a double-precision floating-point number representing the value of S). Two examples are shown. In the first, the p from the last example is converted into a numerical form. In the second, a symbolic object is created, and then converted into numerical form.

```
>> pN=double(p)              p is converted to numerical form (assigned to pN).

pN =
    4.0476

>> y=sym(10)*cos(5*pi/6)            Create a symbolic expression y.

y =                                  Exact value of y is displayed.
-5*3^(1/2)

>> yN=double(y)               y is converted to numerical form (assigned to yN).

yN =
   -8.6603
```

- A symbolic object that is created can also be a symbolic expression written in terms of variables that have not been first created as symbolic objects. For example, the quadratic expression $ax^2 + bx + c$ can be created as a symbolic object named f by using the `sym` command:

```
>> f=sym('a*x^2+b*x+c')

f =
a*x^2 + b*x +c
```

It is important to understand that in this case, the variables a, b, c, and x included in the object do not exist individually as independent symbolic objects (the whole expression is one object). This means that it is impossible to perform symbolic math operations associated with the individual variables in the object. For example, it will not be possible to differentiate f with respect to x. This is different than the way in which the quadratic expression was created in the first example in Section *11.1.2*, where the individual variables are first created as symbolic objects and then used in the quadratic expression.

- Existing symbolic expressions can be used to create new symbolic expressions. This is done by simply using the name of the existing expression in the new expression. For example:

```
>> syms x y              Define x, and y as symbolic variables.
>> SA=x+y, SB=x-y        Create two symbolic expressions SA and SB.
SA =
x+y                                        SA = x + y
SB =
x-y                                        SB = x - y
>> F=SA^2/SB^3+x^2   Create a new symbolic expression F using SA and SB.
F =
(x+y)^2/(x-y)^3+x^2      F = (SA²)/(SB³) + x² = (x+y)²/(x-y)³ + x²
```

11.1.3 The `findsym` Command and the Default Symbolic Variable

The `findsym` command can be used to find which symbolic variables are present in an existing symbolic expression. The format of the command is:

$$\boxed{\text{findsym(S)}} \quad \text{or} \quad \boxed{\text{findsym(S,n)}}$$

The `findsym(S)` command displays the names of all the symbolic variables (separated by commas) that are in the expression S in alphabetical order. The `findsym(S,n)` command displays n symbolic variables which are in expression S in the default order. For one-letter symbolic variables, the default order starts with x and is followed by letters according to their closeness to x. If there are two letters that are equally close to x, the letter that is after x in the alphabetical order is first (y before w, and z before v). The default symbolic variable in a symbolic expression is the first variable in the default order. The default symbolic variable in an expression S can be identified by typing `findsym(S,1)`. Examples:

```
>> syms x h w y d t     Define x, h, w, y, d, and t as symbolic variables.
>> S=h*x^2+d*y^2+t*w^2       Create a symbolic expression S.
S =
h*x^2+d*y^2+t*w^2
>> findsym(S)                       Use the findsym(S) command.
ans =              The symbolic variables are displayed in alphabetical order.
d, h, t, w, x, y
>> findsym(S,5)              Use the findsym(S,n) command (n = 5).
```

```
ans =
x,y,w,t,h
```
5 symbolic variables are displayed in the default order.
```
>> findsym(S,1)
```
Use the `findsym(S,n)` command with n = 1.
```
ans =
x
```
The default symbolic variable is displayed.

11.2 CHANGING THE FORM OF AN EXISTING SYMBOLIC EXPRESSION

Symbolic expressions are either created by the user or by MATLAB as the result of symbolic operations. The expressions created by MATLAB might not be in the simplest form or in a form that the user prefers. The form of an existing symbolic expression can be changed by collecting terms with the same power, by expanding products, by factoring out common multipliers, by using mathematical and trigonometric identities, and by many other operations. The following sub-sections describe several of the commands that can be used to change the form of an existing symbolic expression.

11.2.1 The collect, expand, and factor Commands

The collect, expand, and factor commands can be used to perform the mathematical operations that are implied by their name.

The collect command:

The collect command collects the terms in the expression that have the variable with the same power. In the new expression, the terms will be ordered in decreasing order of power. The command has the forms:

collect(S)   collect(S, variable_name)

where S is the expression. The collect(S) form works best when an expression has only one symbolic variable. If an expression has more than one variable MATLAB will collect the terms of one variable first, then of a second variable, and so on. The order of the variables is determined by MATLAB. The user can specify the first variable by using the collect(S, variable_name) form of the command. Examples:

```
>> syms x y
```
Define x and y as symbolic variables.
```
>> S=(x^2+x-exp(x))*(x+3)
S =
(x^2 + x-exp(x))*(x + 3)
```
Create the symbolic expression S:
$(x^2 + x - e^x)(x + 3)$.
```
>> F = collect(S)
```
Use the collect command.
```
F =
x^3+4*x^2+(-exp(x)+3)*x-3*exp(x)
```
MATLAB returns the expression:
$x^3 + 4x^2 + (- e^x + 3)x - 3e^x$.

```
>> T=(2*x^2+y^2)*(x+y^2+3)
T =
(2*x^2+y^2)*(x+y^2+3)
```
Create the symbolic expression T: $(2x^2 + y^2)(x + y^2 + 3)$.

```
>> G=collect(T)
```
Use the `collect(T)` command.

MATLAB returns the expression: $2x^3 + (2y^2 + 6)x^2 + xy^2 + y^2(y^2 + 3)$.

```
G =
2*x^3+(2*y^2+6)*x^2+x*y^2+y^2*(y^2+3)
```

```
>> H=collect(T,y)
```
Use the `collect(T, y)` command.

```
H =
y^4+(2*x^2+x+3)*y^2+2*x^2*(x+3)
```
MATLAB returns the expression: $y^4 + (2x^2 + x + 3)y^2 + 2x^2(x + 3)$.

Note above that when `collect(T)` is used, the reformatted expression is written in terms of reduced powers of x, but when the `collect(T, y)` is used the reformatted expression is written in terms of reduced powers of y.

The `expand` command:

The `expand` command expands expressions in two ways. It carries out products of terms that include summation (at least one of the terms), and it uses trigonometric identities and exponential and logarithmic laws to expand corresponding terms that include summation. The form of the command is:

$$\boxed{\texttt{expand(S)}}$$

where S is the symbolic expression. Two examples are:

```
>> syms a x y
```
Define a, x, and y as symbolic variables.

```
>> S=(x+5)*(x-a)*(x+4)
S =
(x+5)*(x-a)*(x+4)
```
Create the symbolic expression S: $(x + 5)(x - a)(x + 4)$.

```
>> T=expand(S)
```
Use the `expand` command.

```
T =
x^3+9*x^2-x^2*a-9*x*a+20*x-20*a
```
MATLAB returns the expression: $x^3 + 9x^2 - ax^2 - 9ax + 20x - 20a$.

```
>> expand(sin(x-y))
```
Use the `expand` command to expand $\sin(x - y)$.

```
ans =
sin(x)*cos(y)-cos(x)*sin(y)
```
MATLAB uses trig identity for the expansion.

The `factor` command:

The `factor` command changes an expression that is a polynomial to be a product of polynomials of a lower degree. The form of the command is:

$$\boxed{\texttt{factor(S)}}$$

where `S` is the symbolic expression. An example is:

`>> syms x`	Define x as a symbolic variable.

`>> S=x^3+4*x^2-11*x-30`
`S =`
`x^3+4*x^2-11*x-30`
	Create the symbolic expression `S`:
	$x^3 + 4x^2 - 11x - 30$.

`>> factor(S)`	Use the `factor` command.
`ans =` `(x+2)*(x-3)*(x+5)`	MATLAB returns the expression: $(x + 2)(x - 3)(x + 5)$.

11.2.2 The `simplify` and `simple` Commands

The `simplify` and `simple` commands are both general tools for simplifying the form of an expression. The `simplify` command uses built-in simplification rules to generate a form of the expression with a simpler form than the original expression. The `simple` command is programmed to generate a form of the expression with the least number of characters. Although there is no guarantee that the form with the least number of characters is the simplest, in actuality this is often the case.

The `simplify` command:

The `simplify` command uses mathematical operations (addition, multiplication, rules of fractions, powers, logarithms, etc.) and functional and trigonometric identities to generate a simpler form of the expression. The format of the `simplify` command is:

$$\boxed{\texttt{simplify(S)}}$$

    `S` is the name of the existing   or   An expression to be simplified
    expression to be simplified.   can be typed in for `S`.

Two examples are:

`>> syms x y`	Define x as a symbolic variable.

`>> S=x*(x*(x-8)+10)-5`
`S =`
`x*(x*(x - 8) + 10) - 5`
	Create the symbolic expression:
	$x(x(x - 8) + 10) - 5$), and assign it to `S`.

`>> SA = simplify(S)`	Use the `simplify` command to Simplify `S`.
`SA =` `x^3-8*x^2+10*x-5`	MATLAB simplifies the expression to: $x^3 - 8x^2 + 10x - 5$.
`>> simplify((x+y)/(1/x+1/y))`	Simplify $(x + y)/\left(\dfrac{1}{x} + \dfrac{1}{y}\right)$.
`ans =` `x*y`	MATLAB simplifies the expression to: xy).

The simple **command:**

The simple command finds a form of the expression with the fewest number of characters. In many cases this form is also the simplest. When the command is executed, MATLAB creates several forms of the expression by applying the collect, expand, factor, simplify commands, and other simplification functions that are not covered here. Then MATLAB returns the expression with the shortest form. The simple command has the following three forms:

F = simple(S)	simple(S)	[F how] = simple(S)
The shortest form of S is assigned to F.	All the simplification trails are displayed. The shortest is assigned to ans.	The shortest form of S is assigned to F. The name (string) of the simplification method is assigned to how.

The difference between the forms is in the output. The use of two of the forms is shown next.

```
>> syms x
```
Define x as a symbolic variable.

```
>> S=(x^3-4*x^2+16*x)/(x^3+64)
S =
(x^3-4*x^2+16*x)/(x^3+64)
```
Create the symbolic expression: $\dfrac{x^3 - 4x^2 + 16x}{x^3 + 64}$, and assign it to S.

```
>> F = simple(S)
F =
x/(x+4)
```
Use the F = simple(S) command to Simplify S.

The simplest form of S, $x/(x+4)$ is assigned to F.

```
>> [G how] = simple(S)
G =
x/(x+4)
how =
factor
```
Use the [G how] = simple(S) command.

The simplest form of S, $x/(x+4)$ is assigned to G.

The word factor is assigned to G, which means that the shortest form was obtained using the factor command.

The use of the simple(S) form of the command is not demonstrated because the display of the output is lengthy. MATLAB displays ten different tries and assigns the shortest form to ans. The reader should try to execute the command and examine the output display.

11.2.3 The pretty Command

The pretty command displays a symbolic expression in a format resembling the mathematical format in which expressions are generally typed. The command has the form:

```
pretty(S)
```

For example:

```
>> syms a b c x                    Define a, b, c, and x as symbolic variables.
>> S=sqrt(a*x^2 + b*x + c)          Create the symbolic expression:
S =                                 √(ax² + bx + c), and assign it to S.
(a*x^2+b*x+c)^(1/2)
>> pretty(S)                        The pretty command displays
                                    the expression in a math format.
              2           1/2
        (a x   + b x + c)
```

11.3 SOLVING ALGEBRAIC EQUATIONS

A single algebraic equation can be solved for one variable, and a system of equations can be solved for several variables with the `solve` function.

Solving a single equation:

An algebraic equation can have one or several symbolic variables. If the equation has one variable, the solution is numerical. If the equation has several symbolic variables, a solution can be obtained for any of the variables in terms of the others. The solution is obtained by using the `solve` command which has the forms:

$$h = \text{solve(eq)} \quad \text{or} \quad h = \text{solve(eq,var)}$$

- The argument `eq` can be the name of a previously created symbolic expression, or an expression that is typed in. When a previously created symbolic expression S is entered for `eq`, or when an expression that does not contain the = sign is typed in for `eq`, MATLAB solves the equation `eq` = 0.

- An equation of the form $f(x) = g(x)$ can be solved by typing the equation (including the = sign) as a string for `eq`.

- If the equation to be solved has more than one variable, the `solve(eq)` command solves for the default symbolic variable (Section *11.1.3*). A solution for any of the variables can be obtained with the `solve(eq,var)` command by typing the variable name for `var`.

- If the user types: `solve(eq)`, the solution is assigned to the variable `ans`.

- If the equation has more than one solution, the output `h` is a symbolic column vector with a solution at each element. The elements of the vector are symbolic objects. When an array of symbolic objects is displayed, each row is enclosed with square brackets (see the examples on the following page).

The following examples illustrate the use of the `solve` command.

```
>> syms a b x y z
```
Define a, b, x, y and z as symbolic variables.

```
>> h=solve(exp(2*z)-5)
```
Use the `solve` command to solve $e^{2z} - 5 = 0$.

```
h =
1/2*log(5)
```
The solution is assigned to h.

```
>> S=x^2-x-6
S =
x^2-x-6
```
Create the symbolic expression: $x^2 - x - 6$, and assign it to S.

```
>> k=solve(S)
```
Use the `solve(S)` command to solve $x^2 - x - 6 = 0$.

```
k =
[ -2]
[  3]
```
The equation has two solutions. They are assigned to k, which is a column vector with symbolic objects.

```
>> solve('cos(2*y)+3*sin(y)=2')

ans =
[ 1/2*pi]
[ 1/6*pi]
[ 5/6*pi]
```
Use the `solve` command to solve: $\cos(2y) + 3\sin(y) = 2$. (The equation is typed as a string in the command.)

The solution is assigned to `ans`.

```
>> T= a*x^2+5*b*x+20
T =
a*x^2+5*b*x+20
```
Create the symbolic expression: $ax^2 + 5bx + 20$, and assign it to T.

```
>> solve(T)
```
Use the `solve(S)` command to solve $T = 0$.

```
ans =
[ 1/2/a*(-5*b+(25*b^2-80*a)^(1/2))]
[ 1/2/a*(-5*b-(25*b^2-80*a)^(1/2))]
```
The equation $T = 0$ is solved for the variable x, which is the default variable.

```
>> M = solve(T,a)
```
Use the `solve(eq,var)` command to solve $T = 0$.

```
M =
-5*(b*x+4)/x^2
```
The equation $T = 0$ is solved for the variable a.

- It is also possible to use the `solve` command by typing the equation to be solved as a string, without having the variables in the equation first created as symbolic objects. However, if the solution contains variables (when the equation has more than one variable), the variables do not exist as independent symbolic objects. For example:

```
>> ts=solve('4*t*h^2+20*t-5*g')
```
The expression $4th^2 + 20t - 5g$ is typed in the `solve` command.

```
ts =
5/4*g/(h^2+5)
```
The variables t, h, and g have not been created as symbolic variables before the expression was typed in the `solve` command.

MATLAB solves the equation $4th^2 + 20t - 5g = 0$ for t.

The equation can also be solved for a different variable. For example, a solution for g is obtained by:

```
>> gs=solve('4*t*h^2+20*t-5*g','g')

gs =
4/5*t*h^2+4*t
```

Solving a system of equations:

The `solve` command can also be used for solving a system of equations. If the number of equations and variables is the same, the solution is numerical. If the number of variables is greater than the number of equations, the solution is symbolic for the desired variables in terms of the other variables. A system of equations (depending on the type of equations) can have one or several solutions. If the system has one solution, each of the variables for which the system is solved has one numerical value (or expression). If the system has more than one solution, each of the variables can have several values.

      The format of the `solve` command for solving a system of n equations is:

$$\boxed{\texttt{output = solve(eq1,eq2,....,eqn)}}$$

or

$$\boxed{\texttt{output = solve(eq1,eq2,....,eqn,var1,var2,....varn)}}$$

- The arguments `eq1,eq2,....,eqn` are the equations to be solved. Each argument can be a name of a previously created symbolic expression, or an expression that is typed in as a string. When a previously created symbolic expression S is entered, the equation is S = 0. When a string that does not contain the = sign is typed in, the equation is `expression` = 0. An equation that contains the = sign must be typed as a string.

- In the first format, if the number of equations n is equal to the number of variables in the equations, MATLAB gives a numerical solution for all the variables. If the number of variables is greater than the number of equations n, MATLAB gives a solution for n variables in terms of the rest of the variables. The variables for which solutions are obtained are chosen by MATLAB according to the default order (Section *11.1.3*).

- When the number of variables is greater than the number of equations n, the user can select the variables for which the system is solved. This is done by using the second format of the `solve` command and entering the names of the variables `var1,var2,....varn`.

The `output` from the `solve` command, which is the solution of the system, can have two different forms. One is a cell array and the other is a structure. A cell array is an array in which each of the elements can be an array. A structure is an array in which the elements (called fields) are addressed by textual field designators. The fields of a structure can be arrays of different sizes and types. Cell arrays and structures are not presented in detail in this book, but a short explanation is given below, so the reader will be able to use them with the `solve` command.

When a cell array is used in the output of the `solve` command, the command has the following form (in the case of a system of three equations):

$$[\text{varA, varB, varC}] = \text{solve}(\text{eq1,eq2,eq3})$$

- Once the command is executed, the solution is assigned to the variables `varA`, `vary`, `varC`, and the variables are displayed with their assigned solution. Each of the variables will have one or several values (in a column vector) depending whether the system of equations has one or several solutions.

- The user can select any name for `varA`, `varB`, `varC`. MATLAB assigns the solution for the variables in the equations in alphabetical order. For example, if the variables for which the equations are solved are x, u, and t, the solution for t is assigned to `varA`, the solution for u is assigned to `varB`, and the solution for x is assigned to `varC`.

The following examples show how the `solve` command is used for the case when cell array is used in the output:

```
>> syms x y t                          Define x, y and t as symbolic variables.

>> S=10*x+12*y+16*t;                    Assign to S the expression: 10x + 12y + 16t.

>> [xt yt]=solve(S, '5*x-y=13*t')      Use the solve command to solve
xt =                                    the system:  10x + 12y + 16t = 0
2*t                                                   5x − y = 13t
yt =
-3*t                                    Output in a cell array with two cells named xt and yt.
                                        The solution for x and y is assigned to xt and yt, respectively.
```

In the example above, notice that the system of two equations is solved by MATLAB for x and y in terms of t, since x and y are the first two variables in the default order. The system, however, can be solved for different variables. As an example, the system is solved next for y and t in terms of x (using the second form of the `solve` command:

```
>> [tx yx]=solve(S,'5*x-y=13*t',y,t)
```

The variables for which the system is solved (y and t) are entered.

```
tx =
1/2*x
yx =
-3/2*x
```

The solution for the variables for which the system is solved is assigned in alphabetical order. The first cell has the solution for t, and the second cell has the solution for y.

When a structure is used in the output of the `solve` command, the command has the form (in the case of a system of three equations):

$$AN = solve(eq1, eq2, eq3)$$

- `AN` is the name of the structure.

- Once the command is executed the solution is assigned to `AN`. MATLAB displays the name of the structure and the names of the fields of the structure which are the names of the variables for which the equations are solved. The size and the type of each field is displayed next to the field name. The content of each field, which is the solution for the variable, is not displayed.

- To display the content of a field (the solution for the variable), the user has to type the address of the field. The form for typing the address is: `structure_name.field_name` (see example below).

As an example, the system of equations solved in the last example is solved again using a structure for the output.

```
>> syms x y t
>> S=10*x+12*y+16*t;
>> AN=solve(S,'5*x-y=3*t')
```
Use the `solve` command to solve the system: $10x + 12y + 16t = 0$
$5x - y = 13t$

```
AN =
  x: [1x1 sym]
  y: [1x1 sym]
```
MATLAB displays the name of the structure `AN` and the names of its fields x & y (size and type), which are the names of the variables for which the equations are solved.

```
>> AN.x
ans =
2*t
```
Type the address of the field x.
The content of the field (the solution for x) is displayed.

```
>> AN.y
ans =
-3*t
```
Type the address of the field y.
The content of the field (the solution for y) is displayed.

Sample problem 11-1 shows a solution of a system of equations that have two solutions.

Sample Problem 11-1: Intersection between a circle and a line

The equation of a circle in the x-y plane with a radius R and its center at point (2, 4) is given by: $(x - 2)^2 + (y - 4)^2 = R^2$. The equation of a line in the plane is given by: $y = \frac{x}{2} + 1$. Determine the coordinates of the points (as a function of R) where the line intersects the circle.

Solution

The solution is obtained by solving the system of the two equations for x and y in terms of R. To show the difference in the output between using cell array and structure output forms of the `solve` command, the system is solved twice. The first solution is with the output in a cell array:

```
>> syms x y R
```

The two equations are typed in the `solve` command.

```
>> [xc,yc]=solve('(x-2)^2+(y-4)^2=R^2','y=x/2+1')
```

Output in a cell array.

```
xc =
[ 14/5+2/5*(-16+5*R^2)^(1/2)]
[ 14/5-2/5*(-16+5*R^2)^(1/2)]
yc =
[ 12/5+1/5*(-16+5*R^2)^(1/2)]
[ 12/5-1/5*(-16+5*R^2)^(1/2)]
```

Output in a cell array with two cells named `xc` and `yc`. Each cell contains two solutions in a symbolic column vector.

The second solution is with the output in a structure:

```
>> COORD=solve('(x-2)^2+(y-4)^2=R^2','y = x/2+1')
```

Output in a structure.

```
COORD =
    x: [2x1 sym]
    y: [2x1 sym]
```

Output in a structure named `COORD` that has two fields `x` and `y`. Each field is a 2 by 1 symbolic vector.

```
>> COORD.x
```

Type the address of the field `x`.

```
ans =
[ 14/5+2/5*(-16+5*R^2)^(1/2)]
[ 14/5-2/5*(-16+5*R^2)^(1/2)]
```

The content of the field (the solution for x) is displayed.

```
>> COORD.y
```

Type the address of the field `y`.

```
ans =
[ 12/5+1/5*(-16+5*R^2)^(1/2)]
[ 12/5-1/5*(-16+5*R^2)^(1/2)]
```

The content of the field (the solution for y) is displayed.

11.4 DIFFERENTIATION

Symbolic differentiation can be carried out by using the `diff` command. The form of the command is:

`diff(S)`   or   `diff(S,var)`

- `S` can either be a name of a previously created symbolic expression, or an expression can be typed in for `S`.

- In the `diff(S)` command, if the expression contains one symbolic variable, the differentiation is carried out with respect to that variable. If the expression contains more than one variable, the differentiation is carried out with respect to the default symbolic variable (Section *11.1.3*).

- In the `diff(S,var)` command (which is used for differentiation of expressions with several symbolic variables) the differentiation is carried out with respect to the variable `var`.

- Second or higher (*n*-th) derivative can determined with the `diff(S,n)` or `diff(S,var,n)` command, where n is a positive number. *n* = 2 for the second derivative, 3 for the third, and so on.

Some examples are:

`>> syms x y t`	Define x, y and t as symbolic variables.
`>> S=exp(x^4);`	Assign to S the expression: e^{x^4}.
`>> diff(S)`	Use the `diff(S)` command to differentiate S.
`ans =` `4*x^3*exp(x^4)`	The answer $4x^3 e^{x^4}$ is displayed.
`>> diff((1-4*x)^3)`	Use the `diff(S)` command to differentiate $(1-4x)^3$.
`ans =` `-12*(1-4*x)^2`	The answer $-12(1-4x)^2$ is displayed.
`>> R=5*y^2*cos(3*t);`	Assign to R the expression: $5y^2\cos(3t)$.
`>> diff(R)`	Use the `diff(R)` command to differentiate R.
`ans =` `10*y*cos(3*t)`	MATLAB differentiates R with respect to *y* (default symbolic variable), the answer $10y\cos(3t)$ is displayed.
`>> diff(R,t)`	Use the `diff(R,t)` command to differentiate R w.r.t. *t*.
`ans =` `-15*y^2*sin(3*t)`	The answer $-15y^2\sin(3t)$ is displayed.
`>> diff(S,2)`	Use `diff(S,2)` command to obtain the 2nd derivative of S.
`ans =` `12*x^2*exp(x^4)+16*x^6*exp(x^4)`	The answer $12x^2 e^{x^4} + 16x^6 e^{x^4}$ is displayed.

- It is possible also to use the `diff` command by typing the expression to be differentiated as a string directly in the command without having the variables in the expression first created as symbolic objects. However, the variables in the differentiated expression do not exist as independent symbolic objects.

11.5 INTEGRATION

Symbolic integration can be carried out by using the `int` command. The command can be used for determining indefinite integrals (antiderivatives) and definite integrals. For indefinite integration the form of the command is:

$$\boxed{\texttt{int(S)}} \qquad \text{or} \qquad \boxed{\texttt{int(S,var)}}$$

- `S` can either be a name of a previously created symbolic expression, or an expression can be typed in for `S`.

- In the `int(S)` command, if the expression contains one symbolic variable, the integration is carried out with respect to that variable. If the expression contains more than one variable, the integration is carried out with respect to the default symbolic variable (Section *11.1.3*).

- In the `int(S,var)` command, which is used for integration of expressions with several symbolic variables, the integration is carried out with respect to the variable `var`.

Some examples are:

`>> syms x y t`	Define x, y and t as symbolic variables.
`>> S=2*cos(x)-6*x;`	Assign to S the expression: $2\cos(x) - 6x$.
`>> int(S)`	Use the `int(S)` command to integrate S.
`ans =` `2*sin(x)-3*x^2`	The answer $2\sin(x) - 3x^2$ is displayed.
`>> int(x*sin(x))`	Use the `int(S)` command to integrate $x\sin(x)$.
`ans =` `sin(x)-x*cos(x)`	The answer $\sin(x) - x\cos(x)$ is displayed.
`>>R=5*y^2*cos(4*t);`	Assign to R the expression: $5y^2\cos(4t)$.
`>> int(R)`	Use the `int(R)` command to integrate R.
`ans =` `5/3*y^3*cos(4*t)`	MATLAB integrate R with respect to y (default symbolic variable), the answer $5y^3\cos(4t)/3$ is displayed.
`>> int(R,t)`	Use the `int(R,t)` command to integrate R w.r.t. t.
`ans =` `5/4*y^2*sin(4*t)`	The answer $5y^2\sin(4t)/4$ is displayed.

For definite integration the form of the command is:

$$\boxed{\texttt{int(S,a,b)}} \qquad \text{or} \qquad \boxed{\texttt{int(S,var,a,b)}}$$

- `a` and `b` are the limits of integration. The limits can be numbers or symbolic variables.

For example, determination of the definite integral $\int_0^\pi (\sin y - 5y^2)dy$ with MAT-LAB is:

```
>> syms y
>> int(sin(y)-5*y^2,0,pi)
ans =
2-5/3*pi^3
```

- It is possible also to use the `int` command by typing the expression to be integrated as a string without having the variables in the expression first created as symbolic objects. However, the variables in the integrated expression do not exist as independent symbolic objects.

- Integration can sometimes be a difficult task. A closed form answer may not exist, or if exists, MATLAB might not be able to find it. When that happens MATLAB returns `int(S)` and a message `Explicit integral could not be found`.

11.6 SOLVING AN ORDINARY DIFFERENTIAL EQUATION

An ordinary differential equation (ODE) can be solved symbolically with the `dsolve` command. The command can be used to solve a single equation or a system of equations. Only single equations are addressed here. Chapter 10 discusses using MATLAB to solve first order ODE's numerically. The reader's familiarity with the subject of differential equations is assumed. The purpose of this section is to show how to use MATLAB for solving such equations.

A first order ODE is an equation that contains the derivative of the dependent variable. If t is the independent variable and y is the dependent variable, the equation can be written in the form:

$$\frac{dy}{dt} = f(t, y)$$

A second order ODE contains the second derivative of the dependent variable (it can also contain the first derivative). Its general form is:

$$\frac{d^2y}{dt^2} = f\left(t, y, \frac{dy}{dt}\right)$$

A solution is a function $y = f(t)$ that satisfies the equation. The solution can be general or particular. A general solution contains constants. In a particular solution the constants are determined to have specific numerical values such that the solution satisfies specific initial or boundary conditions.

The command `dsolve` can be used for obtaining a general solution, or, when the initial or boundary conditions are specified, for obtaining a particular solution.

General solution:

For obtaining a general solution, the `dsolve` command has the form:

$$\boxed{\texttt{dsolve('eq')}} \quad \text{or} \quad \boxed{\texttt{dsolve('eq','var')}}$$

- `eq` is the equation to be solved. It has to be typed as a string (even if the variables are symbolic objects).

- The variables in the equation don't have to first be created as symbolic objects. (Then, in the solution the variables will not be symbolic objects.)

- Any letter (lower or upper case), except `D` can be used for the dependent variable.

- In the `dsolve('eq')` command the independent variable is assumed (default) by MATLAB to be `t`.

- In the `dsolve('eq','var')` command the user defines the independent variable by typing it for `var` (as a string).

- When typing the equation the letter `D` denotes differentiation. If y is the dependent variable and t is the independent variable, `Dy` stands for $\dfrac{dy}{dt}$. For example, the equation $\dfrac{dy}{dt} + 3y = 100$ is typed in as: `'Dy + 3*y = 100'`.

- A second derivative is typed by `D2`, third derivative by `D3`, and so on. For example, the equation $\dfrac{d^2y}{dt^2} + 3\dfrac{dy}{dt} + 5y = \sin(t)$ is typed in as: `'D2y + 3*Dy + 5*y = sin(t)'`.

- The variables in the ODE equation that is typed in the `dsolve` command do not have to be previously created symbolic variables.

- In the solution MATLAB uses `C1`, `C2`, `C3`, and so on, for the constants of integration.

For example, a general solution of the first order ODE: $\dfrac{dy}{dt} = 4t + 2y$ is obtained by:

```
>> dsolve('Dy=4*t+2*y')

ans =
-2*t-1+exp(2*t)*C1
```
The answer $y = -2t - 1 + C_1 e^{2t}$ is displayed.

A general solution of the second order ODE: $\dfrac{d^2x}{dt^2} + 2\dfrac{dx}{dt} + x = 0$ is obtained by:

```
>> dsolve('D2x+2*Dx+x=0')

ans =
C1*exp(-t)+C2*exp(-t)*t
```
The answer $x = C_1 e^{-t} + C_2 t e^{-t}$ is displayed.

The following examples illustrate the solution of differential equations that contain symbolic variables in addition to the independent and dependent variables.

```
>> dsolve('Ds=a*x^2')
```
The independent variable is t (default).

MATLAB solves the equation: $\dfrac{ds}{dt} = ax^2$.

```
ans =
a*x^2*t+C1
```
The solution $s = ax^2t + C_1$ is displayed.

```
>> dsolve('Ds=a*x^2','x')
```
The independent variable is defined to be x.

MATLAB solves the equation: $\dfrac{ds}{dx} = ax^2$.

```
ans =
1/3*a*x^3+C1
```
The solution $s = \dfrac{1}{3}ax^3 + C_1$ is displayed.

```
>> dsolve('Ds=a*x^2','a')
```
The independent variable is defined to be a.

MATLAB solves the equation: $\dfrac{ds}{da} = ax^2$.

```
ans =
1/2*a^2*x^2+C1
```
The solution $s = \dfrac{1}{2}a^2x^2 + C_1$ is displayed.

Particular solution:

A particular solution of an ODE can be obtained if boundary (or initial) conditions are specified. A first order equation requires one condition, a second order equation requires two conditions, and so on. For obtaining a particular solution, the dsolve command has the form:

First order ODE:

```
dsolve('eq','cond1','var')
```

Higher order ODE:

```
dsolve('eq','cond1','cond2',....,'var')
```

- For solving equations of higher order, additional boundary conditions have to be entered in the command. If the number of conditions is less than the order of the equation, MATLAB returns a solution that includes constants of integration (C1, C2, C3, and so on).

- The boundary conditions are typed in as strings in the following way:

Math form	MATLAB form
$y(a) = A$	`'y(a)=A'`
$y'(a) = A$	`'Dy(a)=A'`
$y''(a) = A$	`'D2y(a)=A'`

- The argument `'var'` is optional, and is used to define the independent variable in the equation. If none is entered, the default is t.

For example, the first order ODE: $\dfrac{dy}{dt} + 4y = 60$, with the initial condition: $y(0) = 5$ is solved with MATLAB by:

```
>> dsolve('Dy+4*y=60','y(0)=5')
ans =
15-10*exp(-4*t)
```
The answer $y = 15 - 10e^{-4t}$ is displayed.

The second order ODE: $\dfrac{d^2y}{dt^2} - 2\dfrac{dy}{dt} + 2y = 0$, $y(0) = 1$, $\left.\dfrac{dy}{dt}\right|_{t=0} = 0$ can be solved with MATLAB by:

```
>> dsolve('D2y-2*Dy+2*y=0','y(0)=1','Dy(0)=0')

ans =
-exp(t)*sin(t)+exp(t)*cos(t)
>> factor(ans)
ans =
-exp(t)*(sin(t)-cos(t))
```
The answer $y = -e^t \sin(t) + e^t \cos(t)$ is displayed.

The answer can be simplified with the `factor` command.

The simplified answer $y = -e^t(\sin(t) - \cos(t))$ is displayed.

Additional examples of solving differential equations are shown in Sample Problem 11-5.

If MATLAB cannot find a solution, it returns an empty symbolic object, and a message: `Warning: explicit solution could not be found.`

11.7 PLOTTING SYMBOLIC EXPRESSIONS

In many cases, there is a need to plot a symbolic expression. This can easily be done with the `ezplot` command. For a symbolic expression S that contains one variable `var`, MATLAB considers the expression to be a function $S(var)$, and the command creates a plot of $S(var)$ versus *var*. For a symbolic expression that contains two symbolic variables `var1` and `var2`, MATLAB considers the expression to be a function in the form $S(var1, var2) = 0$, and the command creates a plot of one variable versus the other.

To plot a symbolic expression S that contains one or two variables, the `ezplot` command is:

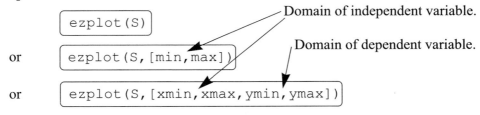

```
ezplot(S)
```
or
```
ezplot(S,[min,max])
```
or
```
ezplot(S,[xmin,xmax,ymin,ymax])
```

Domain of independent variable.

Domain of dependent variable.

- S is the symbolic expression to be plotted. It can be the name of a previously created symbolic expression, or an expression can be typed in for S.

- It is also possible to type the expression to be plotted as a string without having the variables in the expression first created as symbolic objects.

- If S has one symbolic variable, a plot of *S(var)* versus *(var)* is created, with the values of *var* (the independent variable) on the abscissa (horizontal axis), and the values of the *S(var)* on the ordinate (vertical axis).

- If the symbolic expression S has two symbolic variables, var1 and var2, the expression is assumed to be a function with the form *S(var1,var2)* = 0. MATLAB creates a plot of one variable versus the other variable. The variable that is first in alphabetic order is taken to be the independent variable. For example, if the variables in *S* are *x* and *y*, then *x* is the independent variable and is plotted on the abscissa and *y* is the dependent variable plotted on the ordinate. If the variables in *S* are *u* and *v*, then *u* is the independent variable and *v* is the dependent variable.

- In the ezplot(S) command, if S has one variable (*S(var)*), the plot is over the domain: $-2\pi < var < 2\pi$ (default domain) and the range is selected by MATLAB. If S has two variables (*S(var1,var2)*), the plot is over $-2\pi < var1 < 2\pi$ and $-2\pi < var2 < 2\pi$.

- In the ezplot(S,[min,max]) command the domain for the independent variable is defined by min and max: *min < var < max* and the range is selected by MATLAB.

- In the ezplot(S,[xmin,xmax,ymin,ymax]) command the domain for the independent variable is defined by xmin and xmax, and the domain of the dependent variable is defined by ymin and ymax.

The ezplot command can also be used to plot a function that is given in a parametric form. In this case two symbolic expressions, S1 and S2, are involved, where each expression is written in terms of the same symbolic variable (independent parameter). For example, a plot of *y* vs. *x* where $x = x(t)$ and $y = y(t)$. In this case the form of the ezplot command is:

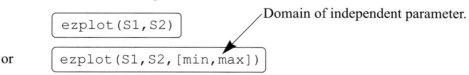

Domain of independent parameter.

```
ezplot(S1,S2)
```

or
```
ezplot(S1,S2,[min,max])
```

- S1 and S2 are symbolic expressions containing the same single symbolic variable, which is the independent parameter. S1 and S2 can be the names of previously created symbolic expressions, or expressions can be typed in.

- The command creates a plot of $S2(var)$ versus $S1(var)$. The symbolic expression that is typed first in the command ($S1$ in the definition above) is used for the horizontal axis and the expression that is typed second ($S2$ in the definition above) is used for the vertical axis.

- In the `ezplot(S1,S2)` command the domain of the independent variable is: $0 < var < 2\pi$ (default domain).

- In the `ezplot(S1,S2,[min,max])` command the domain for the independent variable is defined by `min` and `max`: $min < var < max$.

Additional comments:

Once a plot is created, it can be formatted in the same way as plots created with the `plot` or `fplot` format. This can be done in two ways; by using commands, or by using the Plot Editor (see Section 5.4). When the plot is created, the expression that is plotted is displayed automatically at the top of the plot. MATLAB has additional easy plot functions for plotting two-dimensional polar plots, and for plotting three-dimensional plots. For more information, the reader is referred to the Help menu of the Symbolic Math Toolbox.

Several examples of using the `ezplot` command are shown in Table 11-1.

Table 11-1: Plots with the `ezplot` command

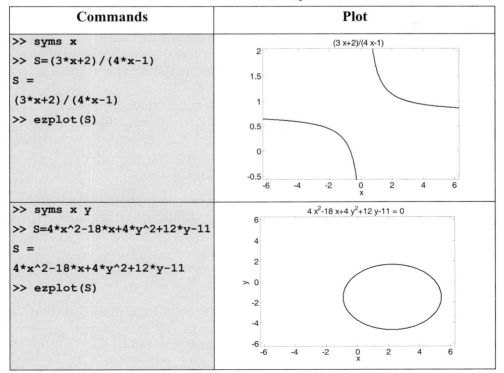

Commands	Plot

`>> syms x`
`>> S=(3*x+2)/(4*x-1)`
`S =`
`(3*x+2)/(4*x-1)`
`>> ezplot(S)`	

`>> syms x y`
`>> S=4*x^2-18*x+4*y^2+12*y-11`
`S =`
`4*x^2-18*x+4*y^2+12*y-11`
`>> ezplot(S)`	

Table 11-1: Plots with the `ezplot` command (Continued)

Commands	Plot

`>> syms t`
`>> x=cos(2*t)`
`x =`
`cos(2*t)`
`>> y=sin(4*t)`
`y =`
`sin(4*t)`
`>> ezplot(x,y)`
	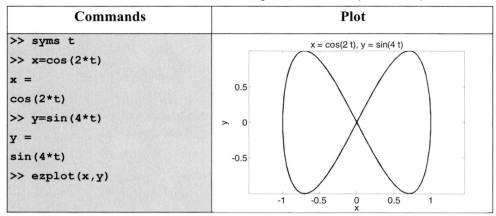

11.8 NUMERICAL CALCULATIONS WITH SYMBOLIC EXPRESSIONS

Once a symbolic expression is created by the user or by the output from any of MATLAB's symbolic operations, there might be a need to substitute numbers for the symbolic variables and calculate the numerical value of the expression. This can be done by using the `subs` command. The `subs` command has several forms and can be used in different ways. The following describes several forms that are easy to use and are suitable for most applications. In one form, the variable (or variables) for which a numerical value is substituted and the numerical value itself are typed inside the `subs` command. In another form, the variable (or variables) is assigned a numerical value in a separate command and then the variable is substituted in the expression.

The `subs` command in which the variable and its value are typed inside the command is shown first. Two cases are presented. One for substituting a numerical value (or values) for one symbolic variable, and the other for substituting numerical values for two or more symbolic variables.

Substituting a numerical value for one symbolic variable:

A numerical value (or values) can be substituted for one symbolic variable when a symbolic expression has one or more symbolic variables. In this case the `subs` command has the form:

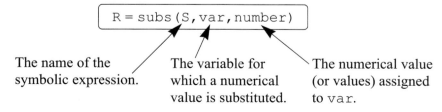

- `number` can be one number (scalar), or an array with many elements (a vector or a matrix).

- The value of S is calculated for each value of number and the result is assigned to R, which will have the same size as number (scalar, vector or matrix).

- If S has one variable, the output R is numerical. If S has several variables and a numerical value is substituted for only one of them, the output R is a symbolic expression.

An example with an expression that includes one symbolic variable is:

```
>> syms x
```
Define x, as a symbolic variable.
```
>> S=0.8*x^3+4*exp(0.5*x)
```
Assign to S the expression: $0.8x^3 + 4e^{(0.5x)}$.
```
S =
4/5*x^3+4*exp(1/2*x)
```
```
>> SD=diff(S)
```
Use the diff(S) command to differentiate S.
```
SD =
12/5*x^2+2*exp(1/2*x)
```
The answer $12x^2/5 + 2e^{(0.5x)}$ is assigned to SD.
```
>> subs(SD, x, 2)
```
Use the subs command to substitute $x = 2$ in SD.
```
ans =
    15.0366
```
The value of SD is displayed.
```
>> SDU=subs(SD, x, [2:0.5:4])
```
Use the subs command to substitute $x = [2, 2.5, 3, 3.5, 4]$ (vector) in SD.
```
SDU =
    15.0366   21.9807   30.5634   40.9092   53.1781
```
The values of SD (assigned to SDU) for each value of x are displayed in a vector.

In the last example, notice that when the numerical value of the symbolic expression is calculated, the answer is numerical (the display is indented). An example of substituting numerical values for one symbolic variable in an expression that has several symbolic variables is:

```
>> syms a g t v
```
Define a, g, t and v as symbolic variables.
```
>> Y=v^2*exp(a*t)/g
```
```
Y =
v^2*exp(a*t)/g
```
Create the symbolic expression: $v^2e^{(at)}/g$ and assigned it to Y.
```
>> subs(Y,t,2)
```
Use the subs command to substitute $t = 2$ in SD.
```
ans =
v^2*exp(2*a)/g
```
The answer $v^2e^{(2a)}/g$ is displayed.
```
>> Yt=subs(Y,t,[2:4])
```
Use the subs command to substitute $t = [2, 3, 4]$ (vector) in Y.

```
Yt =
[ v^2*exp(2*a)/g, v^2*exp(3*a)/g, v^2*exp(4*a)/g]
```
The answer is a vector with elements of symbolic expressions for each value of *t*.

Substituting a numerical value for two or more symbolic variables:

A numerical value (or values) can be substituted for two or more symbolic variables when a symbolic expression has several symbolic variables. In this case the subs command has the form (it is shown for two variables, but can be used in the same form for more):

$$R = subs(S, \{var1, var2\}, \{number1, number2\})$$

The name of the symbolic expression. The variables for which numerical values are substituted. The numerical value (or values) assigned to var1 and var2.

- The variables var1 and var2 are the variables in the expression S for which the numerical values are substituted. The variables are typed as a cell array (inside curly braces { }). Cell array is an array of cells where each cell can be an array of numbers or text.

- The numbers number1, number2 substituted for the variables are also typed as a cell array (inside curly braces { }). The numbers can be scalars, vectors, or matrices. The first cell in the numbers cell array (number1) is substituted for the variable that is in the first cell of the variable cell array (var1), and so on.

- If all the numbers that are substituted for variables are scalars, the outcome will be one number or one expression (if some of the variables are still symbolic).

- If, for at least one variable, the substituted numbers are an array, the mathematical operations are executed element-by-element and the outcome is an array of numbers or expressions. It should be emphasized that the calculations are performed element-by-element even though the expression S is not typed in the element-by-element notation. This also means that all the arrays substituted for different variables must be of the same size.

- It is possible to substitute arrays (of the same size) for some of the variables and scalars for other variables. In this case, in order to carry out element-by-element operations, MATLAB expands the scalars (array of ones times the scalar) to produce an array result.

The substitution of numerical values for two or more variables is demonstrated in the next examples.

```
>> syms a b c e x
```
Define a, b, c, e, and x as symbolic variables.

```
>> S=a*x^e+b*x+c
S =
a*x^e+b*x+c
```
Create the symbolic expression: $ax^e + bx + c$ and assigned it to S.

```
>> subs(S,{a,b,c,e,x},{5,4,-20,2,3})
```
Cell array. Cell array.

Substitute in S scalars for all the symbolic variables.

```
ans =
    37
```
The value of S is displayed.

```
>> T=subs(S,{a,b,c},{6,5,7})
T =
6*x^e+5*x+7
```
Substitute in S scalars for the symbolic variables a, b, and c.

The result is an expression with the variables x, and e.

```
>> R=subs(S,{b,c,e},{[2 4 6],9,[1 3 5]})
```
Substitute in S a scalar for c, and vectors for b, and e.

```
R =
[    a*x+2*x+9,  a*x^3+4*x+9,  a*x^5+6*x+9]
```
The result is a vector of symbolic expressions.

```
>> W=subs(S,{a,b,c,e,x},{[4 2 0],[2 4 6],[2 2 2],[1 3 5],[3 2 1]})
```
Substitute in S vectors for all the variables.

```
W =
    20     26      8
```
The result is a vector of numerical values.

A second method to substitute numerical values for symbolic variables in a symbolic expression is first to assign numerical values to the variables, then to use the subs command. In this method, after the symbolic expression exists (at this point the variables in the expression are symbolic) the variables are assigned numerical values. Then, the subs command is used in the form:

R = subs(S)   The name of the symbolic expression.

Once the symbolic variables are redefined as numerical variables they can no longer be used as symbolic. The method is demonstrated in the following examples.

```
>> syms A c m x y
```
Define A, c, m, x and y as symbolic variables.

```
>> S=A*cos(m*x)+c*y
S =
A*cos(m*x)+c*y
```
Create the symbolic expression: $A\cos(mx) + cy$ and assign it to S.

```
>> A=10; m=0.5; c=3;
```
Assign numerical values to variables A, m, and c.

```
>> subs(S)
```
Use the subs command with the expression S.

```
ans =
10*cos(1/2*x) + 3*y
```
The numerical values of variables A, m, and c are substituted in S.

```
>> x=linspace(0,2*pi,4);
```
Assign numerical values (vector) to variable x.

```
>> T = subs(S)
```
Use the subs command with the expression S.

```
T =
[ 10+3*y, 5+3*y, -5+3*y, -10+3*y]
```
The numerical values of variables A, m, c, and x are substituted. The result is a vector of symbolic expressions.

11.9 EXAMPLES OF MATLAB APPLICATIONS

Sample Problem 11-2: Firing angle of a projectile

A projectile is fired at a speed of 210 m/s and an angle θ. The projectile's intended target is 2600 m away and 210 m above the firing point.

a) Derive the equation that has to be solved in order to determine the angle θ such that the projectile will hit the target.

b) Use MATLAB to solve the equation derived in part *a*.

c) For the angle determined in part *b*, use the ezplot command to make a plot of the projectile's trajectory.

Solution

a) The motion of the projectile can be analyzed by considering the horizontal and vertical components. The initial velocity v_0 can be resolved into horizontal and vertical components:

$$v_{0x} = v_0 \cos(\theta) \quad \text{and} \quad v_{0y} = v_0 \sin(\theta)$$

In the horizontal direction the velocity is constant, and the position of the projectile as a function of time is given by:

$$x = v_{0x}t$$

Substituting $x = 2600$ m for the horizontal distance that the projectile travels to the target, and $210\cos(\theta)$ for v_{0x}, and solving for t gives:

$$t = \frac{2600}{210\cos(\theta)}$$

In the vertical direction the position of the projectile is given by:

$$y = v_{0y}t - \frac{1}{2}gt^2$$

Substituting $y = 350$ m for the vertical coordinate of the target, $210\sin(\theta)$ for v_{0x}, $g = 9.81$, and t gives:

$$350 = 210\sin(\theta)\frac{2600}{210\cos(\theta)} - \frac{1}{2}9.81\left(\frac{2600}{210\cos(\theta)}\right)^2$$

or:

$$350 = 2600\tan(\theta) - \frac{1}{2}9.81\left(\frac{2600}{210\cos(\theta)}\right)^2$$

The solution of this equation gives the angle θ at which the projectile has to be fired.

b) A solution of the equation derived in part a obtained by using the `solve` command (in the Command Window) is:

```
>> syms theta
>>Angle=solve('2600*tan(theta)-0.5*9.81*(2600/(210*cos(theta)))^2=350')
Angle =
[ -2.6823398465577220256847788629067]
[ -1.8962381563523770701488298026235]
[  1.2453544972374161683138135806560]
[   .4592528070320712127786452037279]
```

MATLAB displays four solutions. The two positive ones are relevant to the problem.

```
>> Angle1 = Angle(3)*180/pi
```

Converting the solution in the third element of `Angle` from radians to degrees.

```
Angle1 =
224.1638095027349102964844451808/pi
```

MATLAB displays the answer as a symbolic object in terms of π.

```
>> Angle1=double(Angle1)
Angle1 =
   71.3536
```

Use the `double` command to obtain numerical values for `Angle1`.

```
>> Angle2=Angle(4)*180/pi
```

Converting the solution in the fourth element of `Angle` from radians to degrees.

```
Angle2 =
82.66550526577281830001563667102/pi
```

MATLAB displays the answer as a symbolic object in terms of π.

```
>> Angle2=double(Angle2)
Angle2 =
   26.3132
```

Use the `double` command to obtain numerical values for `Angle2`.

c) The solution from part b shows that there are two possible angles and thus two trajectories. In order to make a plot of a trajectory, the x and y coordinates of the projectile are written in terms of t (parametric form):

$$x = v_0\cos(\theta)t \quad \text{and} \quad y = v_0\sin(\theta)t - \frac{1}{2}gt^2$$

The domain for t is from $t = 0$ until $t = \dfrac{2600}{210\cos(\theta)}$.

These equations can be used in the `ezplot` command to make the plots as shown in the following program written in a script file.

```
xmax=2600; v0=210; g=9.81;
theta1=1.24535; theta2=.45925;
t1=xmax/(v0*cos(theta1));
t2=xmax/(v0*cos(theta2));
syms t
X1=v0*cos(theta1)*t;
X2=v0*cos(theta2)*t;
Y1=v0*sin(theta1)*t-0.5*g*t^2;
Y2=v0*sin(theta2)*t-0.5*g*t^2;
ezplot(X1,Y1,[0,t1])
hold on
ezplot(X2,Y2,[0,t2])
hold off
```

> Assign the two solutions from part *b* to theta1 and theta2.

> Plot one trajectory.

> Plot second trajectory.

When this program is executed the following plot is generated in the Figure Window:

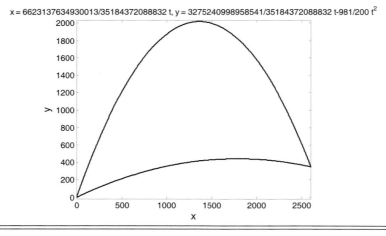

Sample Problem 11-3: Bending resistance of a beam

The bending resistance of a rectangular beam of width b and height h is proportional to the beam's moment of inertia I defined by $I = \frac{1}{12}bh^3$. A rectangular beam is cut out of a cylindrical log of radius R. Determine b and h (as a function of R) such that the beam will have maximum I.

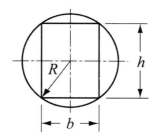

Solution

The problem is solved by following these steps:

a. Write an equation that relates R, h, and b.
b. Derive an expression for I in terms of h.
c. Take the derivative of I with respect to h.
d. Set the derivative equal to zero and solve for h.
e. Determine the corresponding b.

The first step is written by looking at the triangle in the figure. The relationship between R, h, and b is given by the Pythagorean theorem as $\left(\dfrac{b}{2}\right)^2 + \left(\dfrac{h}{2}\right)^2 = R^2$.

Solving this equation for b gives $b = \sqrt{4R^2 - h^2}$.
The rest of the steps are done using MATLAB:

```
>> syms b h R
>> b=sqrt(4*R^2-h^2);                    Create a symbolic expression for b.
>> I=b*h^3/12                            Step b: Create a symbolic expression for I.

I =
1/12*(4*R^2-h^2)^(1/2)*h^3               MATLAB substitutes b in I.

>> ID=diff(I,h)                          Step c: Use the diff(R) command
                                         to differentiate I with respect to h.
ID =
-1/12/(4*R^2-h^2)^(1/2)*h^4+1/4*(4*R^2-h^2)^(1/2)*h^2

                                         The derivative of I is displayed.

>> hs=solve(ID,h)                        Step d: Use the solve command to solve the
                                         equation ID = 0 for h. Assign the answer is to hs.
hs =
[          0]                            MATLAB displays four solutions. The positive
[          0]                            non-zero solution √3R is relevant to the problem.
[   3^(1/2)*R]
[  -3^(1/2)*R]

>> bs=subs(b,hs(3))                      Step e: Use the subs command to determine b by
                                         substituting the solution for h in the expression for b.

bs =                                     The answer for b is displayed. (The answer
(R^2)^(1/2)                              is R, but MATLAB displays (R²)^(1/2).)

>> bss=simple(bs)                        Use the simple command to simplify bs.

bss =                                    The simplified answer for bs is displayed.
R
```

Sample Problem 11-4: Fuel level in a tank

The horizontal cylindrical tank shown is used to store fuel. The tank has a diameter of 6 m and is 8 m long. The amount of fuel in the tank can be estimated by looking at the level of the fuel through a narrow vertical glass window at the front of the tank. A scale that is marked next to the window shows the level of the fuel corresponding to 40, 60, 80, 120, and 160 thousand litters. Determine the vertical position (measured from the ground) of the lines of the scale.

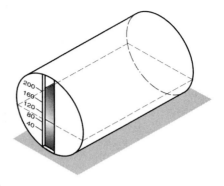

Solution

The relationship between the level of the fuel and its volume can written in the form of a definite integral. Once the integration is carried out, an equation is obtained for the volume in terms of the fuel's height. The height corresponding to a specific volume can then be determined from solving the equation for the height.

The volume of the fuel V can be determined by multiplying the area of the cross section of the fuel A (the shaded area) by the length of the tank L. The cross sectional area can be calculated by integration.

$$V = AL = L \int_0^h w\,dy$$

The width w of the top surface of the fuel can be written as a function of y. From the triangle in the figure on the right, the variables y, w, and R are related by:

$$\left(\frac{w}{2}\right)^2 + (R-y)^2 = R^2$$

Solving this equation for w gives:

$$w = 2\sqrt{R^2 - (R-y)^2}$$

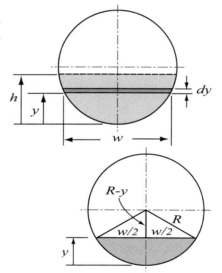

The volume of the fuel at height h can now be calculated by substituting w in the integral in the equation for the volume and carrying out the integration. The result is an equation that gives the volume V as a function of h. The value of h for a given V is obtained by solving the equation for h. In the present problem values of h have to be determined for volumes of 40, 60, 80, 120, and 160 thousand liters. The solution is given in the following MATLAB program (script file):

```
R=3;  L=8;
syms  w  y  h
w=2*sqrt(R^2-(R-y)^2)          Create a symbolic expression for w.
S = L*w                        Create the expression that will be integrated.
V = int(S,y,0,h)               Use the int command to integrate S from 0
                               to h. The result gives V as a function of h.
Vscale=[40:40:200]             Create a vector with the values of V in the scale.
for i=1:5                      Each pass in the loop solves h for one value of V.
    Veq=V-Vscale(i);           Create the equation for h that has to be solved.
    h_ans(i)=solve(Veq);       Use the solve command to solve for h.
end                            h_ans is a vector (symbolic with numbers) with the values
                               of h that correspond to the values of V in the vector Vscale.
h_scale=double(h_ans)          Use the double command to obtain numeri-
                               cal values for the elements of vector h_ans.
```

When the script file is executed, the outcome from commands that don't have a semicolon at the end are displayed. The display in the Command Window is:

```
>> w =
2*(6*y-y^2)^(1/2)              The symbolic expression for w is displayed.

S =
16*(6*y-y^2)^(1/2)            S is the expression that will be integrated.

V =
8*(6*h-h^2)^(1/2)*h-24*(6*h-h^2)^(1/2)+72*asin(-1+1/
3*h)+36*pi
                              The result from the integration; V as a function of h.

Vscale =
    40    80   120   160   200     The values of V in the scale are displayed.

h_scale =
   1.3972   2.3042   3.1439   3.9957   4.9608
                              The position of the lines in the scale are displayed.
```

Units: The unit for length in the solution is meter, which correspond to m^3 for the volume (1 m^3 = 1,000 L).

Sample Problem 11-5: Amount of medication in the body

The amount M of medication present in the body depends on the rate that the medication is consumed by the body and on the rate that the medication enters the body, where the rate at which the medication is consumed is proportional to the amount present in the body. A differential equation for M is:

$$\frac{dM}{dt} = -kM + p$$

where k is the proportionality constant and p is the rate that the medication is injected into the body.

a)   Determine k if the half-life of the medication is 3 hours.

b)   A patient is admitted to a hospital and the medication is given to him at a rate of 50 mg per hour. (Initially there is no medication in the patient's body.) Derive an expression for M as a function of time.

c)   Plot M as a function of time for the first 24 hours.

Solution

a)   The proportionality constant can be determined from considering the case in which the medication is consumed by the body and no new medication is given. In this case the differential equation is:

$$\frac{dM}{dt} = -kM$$

The equation can be solved with the initial condition $M = M_0$ at $t = 0$:

```
>> syms M M0 k t
>> Mt=dsolve('DM=-k*M','M(0)=M0')
Mt =
M0*exp(-k*t)
```
Use the `dsolve` command to solve $\frac{dM}{dt} = -kM$.

The solution gives M as a function of time:

$$M(t) = M_0 e^{-kt}$$

Half-life of 3 hours means that at $t = 3$ hours $M(t) = \frac{1}{2}M_0$. Substituting this information in the solution gives $0.5 = e^{-3k}$, and the constant k is determined from solving this equation:

```
ks=solve('0.5=exp(-k*3)')
ks =
.23104906018664843647241070715273
```
Use the `solve` command to solve $0.5 = e^{-3k}$.

b)   For this part the differential equation for M is:

$$\frac{dM}{dt} = -kM + p$$

The constant k is known from part a, and $p = 50$ mg/h is given. The initial condition is that in the beginning there is no medication in the patient's body, or $M = 0$ at $t = 0$. The solution of this equation with MATLAB is:

```
>> syms p
>> Mtb=dsolve('DM=-k*M+p','M(0)=0')
Mtb =
p/k-p/k*exp(-k*t)
```

> Use the `dsolve` command to solve $\dfrac{dM}{dt} = -kM + p$.

c) A plot of `Mtb` as a function of time for $0 \le t \le 24$ can be done by using the `ezplot` command:

```
>> pgiven=50;
>> Mtt=subs(Mtb,{p,k},{pgiven,ks})
Mtt =
216.404-216.404*exp(-.231049*t)
>> ezplot(Mtt,[0,24])
```

> Substitute numerical values for p and k.

In the actual display of the last expression that was generated by MATLAB (`Mtt = ....`) the numbers have many more decimal digits than shown above. The numbers were shortened so that they will fit in the page.
The plot that is generated is:

216.40425613334451110398870215028-216.40425613334451110398870215028 exp(-.23104906018664843647241070715273 t)

11.10 PROBLEMS

1. Define x as a symbolic variable and create the two symbolic expressions:
$$S_1 = (x-4)^2 - (x+3)^2 + 16x - 4 \text{ and } S_2 = x^3 - 6x^2 - x + 30$$
Use symbolic operations to determine the simplest form of following expressions:

a) $S_1 \cdot S_2$.

b) $\dfrac{S_1}{S_2}$.

c) $S_1 + S_2$.

d) Use the `subs` command to evaluate the numerical value of the result from part c for $x = 2$.

2. Define y as a symbolic variable and create the two symbolic expressions:
$$S_1 = (\sqrt{3} + x)^2 - 2\left(\sqrt{3}x + \frac{x}{2} + \frac{x^2}{2}\right) \text{ and } S_2 = x^2 + 3x + 9$$
Use symbolic operations to determine the simplest form of following expressions:

a) $S_1 \cdot S_2$.

b) $\dfrac{S_1}{S_2}$.

c) $S_1 + S_2$.

d) Use the `subs` command to evaluate the numerical value of the result from part c for $x = 4$.

3. Define u as a symbolic variable and create the two symbolic expressions:
$$Q = u^3 + 2u^2 - 25u - 50 \text{ and } R = 3u^3 + 4u^2 - 75u - 100$$
Use symbolic operations to determine the simplest form of the product Q/R.

4. Define x as a symbolic variable.

a) Show that the roots of the polynomial:
$$f(x) = x^5 - x^4 - 27x^3 + 13x^2 + 134x - 120$$
are 1, 2, 5, –3, and –4 by using the `factor` command.

b) Derive the equation of the polynomial that has the roots: $x = 5$, $x = -3$, $x = -2$, and $x = 4$.

5. Use the commands from Section 11.2 to show that:

a) $\cos(3x) = 4\cos^3 x - 3\cos x$.

b) $\sin x \cos y = \dfrac{1}{2}[\sin(x - y) + \sin(x + y)]$

c) $\begin{aligned}\cos(x + y + z) = {} & \cos x \cos y \cos z - \sin x \sin y \cos z \\ & - \sin x \cos y \sin z - \cos x \sin y \sin z\end{aligned}$

6. The folium of Descartes is the graph shown in the figure. In parametric form its equation is given by:

$$x = \frac{3t}{1 + t^3} \quad \text{and} \quad y = \frac{3t^2}{1 + t^3} \quad \text{for} \quad t \neq -1$$

a) Use MATLAB to show that the equation of the folium of Descartes can also be written as:

$$x^3 + y^3 = 3xy$$

b) Make a plot of the folium, for the domain shown in the figure by using the `ezplot` command.

7. A cylindrical silo with a height of $h = 8\,\text{m}$ and a roof shaped as a cone with a height of $2h$ has a surface area of $370\,\text{m}^2$. Determine the radius R of the base. (Write an equation for the surface area in terms of the radius and the height. Solve the equation for the radius, and use the `double` command to obtain a numerical value.)

8. The relation between the tension T and the steady shortening velocity v in a muscle is given by the Hill equation:

$$(T + a)(v + b) = (T_0 + a)b$$

where a and b are positive constants and T_0 is the isometric tension, i.e. the tension in the muscle when $v = 0$. The maximum shortening velocity occurs when $T = 0$.

a) Using symbolic operations, create the Hill equation as a symbolic expression. Then use `subs` to substitute $T = 0$, and finally solve for v to show that $v_{max} = (bT_0)/a$.

b) Use v_{max} from part a to eliminate the constant b from the Hill equation, and show that $v = \dfrac{a(T_0 - T)}{T_0(T + a)} v_{max}$.

9. Consider the two circles in the x-y plane given by the equations:

$$(x - 2)^2 + (y - 3)^2 = 16 \quad \text{and} \quad x^2 + y^2 = 25$$

a) Use the `ezplot` command to plot the two circles in the same plot.

b) Determine the coordinates of the points where the circles intersect.

10. A 4 ft long boom supports a weight W positioned a distance x from support A as shown. The tension T in the cable and the x and y components of the force at A (F_{Ax} and F_{Ay}) can be calculated from the equations:

$$\frac{12}{5}T - Wx = 0$$

$$F_{Ax} - \frac{4}{5}T = 0$$

$$F_{Ay} + \frac{3}{5}T - W = 0$$

a) Use MATLAB to derive expressions for the forces T, F_{Ax}, and F_{Ay} in terms of x, and W.
b) Use the `subs` command to substitute $W = 200\,\text{lb}$ into the expressions that were derived in part a. This will give the forces as a function of the distance x.
c) Use the `ezplot` command to plot the forces (all three in the same plot) as a function of x, for x starting at 0 and ending at 4 ft.

11. The mechanical power output P in a contracting muscle is given by:

$$P = Tv = \frac{kvT_0\left(1 - \dfrac{v}{v_{max}}\right)}{k + \dfrac{v}{v_{max}}}$$

where T is the muscle tension, v is the shortening velocity (max of v_{max}), T_0 is the isometric tension (i.e. tension at zero velocity) and k is a non-dimensional constant that ranges between 0.15 and 0.25 for most muscles. The equation can be written in non-dimensional form:

$$p = \frac{ku(1 - u)}{k + u}$$

where $p = (Tv)/(T_0 v_{max})$, and $u = v/v_{max}$. Consider the case $k = 0.25$.
a) Plot p versus u for $0 \le u \le 1$.
b) Use differentiation to find the value of u where p is maximum.
c) Find the maximum value of p.

12. The equation of an ellipse is: $\dfrac{x^2}{a^2} + \dfrac{y^2}{b^2} = 1$
where $2a$ and $2b$ are the lengths of the major and minor axis, respectively. Write a program in a a script file that first derives the equation (symbolically) of the tangent line to the ellipse at the point (x_0, y_0) on the upper part of the ellipse (i.e. for $-a < x_0 < a$ and $0 < y_0$). Then, for specific values of a, b, x_0, and y_0 the program makes a plot, like the one shown on the right, of the ellipse and the tangent line. Execute the program with $a = 10$, $b = 7$, and $x_0 = 8$.

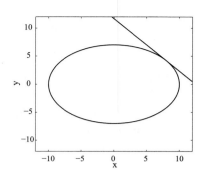

13. A tracking radar is locked on an airplane flying at a constant altitude of 5 km, and a constant speed of 540 km/h. The airplane travels along a path that passes exactly above the radar. The radar starts the tracking when the airplane is 100 km away.

   a) Derive an expression for the angle θ of the radar antenna as a function of time.

   b) Derive an expression for the angular velocity of the antenna, $\dfrac{d\theta}{dt}$, as a function of time.

   c) Make two plots on the same page, one of θ vs. time and the other of $\dfrac{d\theta}{dt}$ vs. time, where the angle is in degrees and the time is in minutes for $0 \le t \le 20$ min.

14. Evaluate the indefinite integral $I = \displaystyle\int \dfrac{\sin^2 x \cos x}{(2 + 3\sin x)^2} \, dx$.

15. Show that the differential volume element of the cone shown is given by:
$$dV = \pi R^2 \left(1 - \dfrac{y}{H}\right)^2 dy$$
Use MATLAB to evaluate the integral of dV from 0 to H symbolically and show that the volume of the cone is $V = \dfrac{1}{3}\pi R^3 H$.

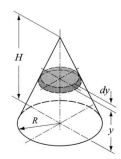

16. The equation of an ellipse is:

$$\frac{x^2}{a^2} + \frac{y^2}{b^2} = 1$$

    Show that the area A enclosed by the ellipse is given by $A = \pi ab$.

17. A ceramic tile has the design shown in the figure. The shaded area is painted red and the rest of the tile is white. The border line between the red and the white areas follows the equation:

    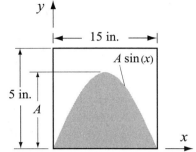

$$y = A\sin(x)$$

    Determine k such that the area of the white and red colors will be the same.

18. Show that the location of the centroid y_c of the cross-sectional area shown is given by $y_c = \frac{H\pi}{8}$. The coordinate y_c can be calculated by:

    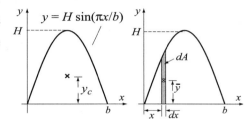

$$y_c = \frac{\int \bar{y}dA}{\int dA}$$

19. The *rms* value of an AC voltage is defined by:

$$v_{rms} = \sqrt{\frac{1}{T}\int_0^T v^2(t')dt'}$$

    where T is the period of the waveform.

    a) A voltage is given by $v(t) = V\cos(\omega t)$. Show that $v_{rms} = \frac{V}{\sqrt{2}}$ and is independent of ω. (The relationship between the period T and the radian frequency ω is: $T = \frac{2\pi}{\omega}$.)

    b) A voltage is given by $v(t) = 2.5\cos(350t) + 3$ V. Determine v_{rms} .

20. The spread of an infection from a single individual to a population of N uninfected persons can be described by the equation:

$$\frac{dx}{dt} = -Rx(N+1-x) \quad \text{with initial condition } x(0) = N$$

    where x is the number of uninfected individuals and R is a positive rate constant. Solve this differential equation symbolically for $x(t)$. Also, determine symbolically the time t at which the infection rate dx/dt is maximum.

21. A resistor R $(R = 0.4\Omega)$ and an inductor L $(L = 0.08\,\mathrm{H})$ are connected as shown. Initially, the switch is connected to point A and there is no current in the circuit. At $t = 0$ the switch is moved from A to B, such that the resistor and the inductor are connected to v_S $(v_S = 6\,\mathrm{V})$, and current starts flowing in the circuit. The switch remains connected to B until the voltage on the resistor reaches 5V. At that time (t_{BA}) the switch is moved back to A.

The current i in the circuit can be calculated from solving the differential equations:

$iR + L\dfrac{di}{dt} = v_S$ During the time from $t = 0$ and until the time when the switch is moved back to A.

$iR + L\dfrac{di}{dt} = 0$ From the time when the switch is moved back to A and on.

The voltage across the resistor v_R at any time is given by $v_R = iR$.

a) Derive an expression for the current i in terms of R, L, v_S, and t for $0 \le t \le t_{BA}$ by solving the first differential equation.

b) Substitute the values of R, L, and v_S in the solution of i, and determine the time t_{BA} when the voltage across the resistor reaches 5V.

c) Derive an expression for the current i in terms of R, L, and t, for $t_{BA} \le t$ by solving the second differential equation.

d) Make two plots (on the same page), one for v_R vs. t for $0 \le t \le t_{BA}$, and the other for v_R vs. t for $t_{BA} \le t \le 2t_{BA}$.

22. The velocity of a sky-diver when his parachute is still closed can be modeled by assuming that the air resistance is proportional to the velocity. From Newton's second law of motion the relationship between the mass m of the sky-diver and his velocity v is given by (down is positive):

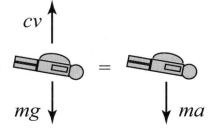

$$mg - cv = m\dfrac{dv}{dt}$$

where c is a drag constant and g is the gravitational constant $g = 9.81 \text{ m/s}^2$.

a) Solve the equation for v in terms of m, g, c, and t, assuming that the initial velocity of the sky-diver is zero.

b) It is observed that 4 s after a 90 kg sky-diver jumps out of an airplane, his velocity is 28 m/s. Determine the constant c.

c) Make a plot of the sky-diver velocity as a function of time for $0 \le t \le 30 \text{ s}$.

23. Determine the general solution of the differential equation:

$$x^2 \frac{d^2y}{dx^2} + 3x\frac{dy}{dx} - 1 = 0$$

Show that the solution is correct. (Derive the first and second derivatives of the solution, and then substitute back in the equation.)

24. Determine the solution of the following differential equation that satisfies the given initial conditions. Plot the solution for $0 \le x \le 5$.

$$\frac{d^3y}{dx^3} + 5\frac{dy}{dx} + 0.5x = 0, \quad y(0) = 0, \quad \left.\frac{dy}{dx}\right|_{x=0} = 0, \quad \left.\frac{d^2y}{dx^2}\right|_{x=0} = 1$$

25. Damped free vibrations can be modeled by considering a block of mass m that is attached to a spring and a dashpot as shown. From Newton's second law of motion, the displacement x of the mass as a function of time can be determined by solving the differential equation:

$$m\frac{d^2x}{dt^2} + c\frac{dx}{dt} + kx = 0$$

where k is the spring constant, and c is the damping coefficient of the dashpot. If the mass is displaced from its equilibrium position and then released, it will start oscillating back and forth. The nature of the oscillations depends on the size of the mass and the values of k and c.

For the system shown in the figure $m = 10 \text{ kg}$, and $k = 28 \text{ N/m}$. At time $t = 0$ the mass is displaced to $x = 0.18 \text{ m}$, and then released from rest. Derive expressions for the displacement x and the velocity v of the mass, as a function of time. Consider the following two cases:

a) $c = 3 \text{ N-s/m}$.

b) $c = 50 \text{ N-s/m}$.

For each case, plot the position x and the velocity v vs. time (two plots on one page). For case (a) take $0 \le t \le 20$ s, and for case (b) take $0 \le t \le 10$ s.

Appendix:
Summary of Characters,
Commands, and Functions

The following tables list MATLAB's characters, commands, and functions that are covered in the book. The items are grouped by subjects.

Characters and arithmetic operators

Character	Description	Page
+	Addition.	11, 58
−	Subtraction.	11, 58
*	Scalar and array multiplication.	11, 59
.*	Element-by-element multiplication of arrays.	66
/	Right division.	11, 65
\	Left division.	10, 64
./	Element-by-element right division.	66
.\	Element-by-element left division.	66
^	Exponentiation.	11
.^	Element-by-element exponentiation.	66
:	Colon; creates vectors with equally spaced elements, represents range of elements in arrays.	35, 41
=	Assignment operator.	16
( )	Parentheses; sets precedence, encloses input arguments in functions and subscripts of arrays.	11, 40, 41, 158
[ ]	Brackets; forms arrays. encloses output arguments in functions.	34, 35, 37, 158
,	Comma; separates array subscripts and function arguments, separates commands in the same line.	9, 17, 40-43, 158
;	Semicolon; suppresses display, ends row in array.	10, 37
'	Single quote; matrix transpose, creates string.	39, 50-52
...	Ellipsis; continuation of line.	9
%	Percent; denotes a comment, specifies output format.	10

Relational and logical operators

Character	Description	Page
<	Less than.	192
>	Greater than.	192
<=	Less than or equal.	192
>=	Greater than or equal.	192
==	Equal.	192
~=	Not equal.	192
&	Logical AND.	195
		Logical OR.	195
~	Logical NOT.	195

Managing commands

Command	Description	Page
cd	Changes current directory.	23
clc	Clears the Command Window.	10
clear	Removes all variables from the memory.	19
clear x y z	Removes variables x, y, and z from the memory.	19
close	Closes the active Figure Window.	144
fclose	Closes a file.	99
figure	Opens a Figure Window.	144
fopen	Opens a file.	98
global	Declares global variables.	161
help	Displays help for MATLAB functions.	160
iskeyword	Displays keywords.	19
lookfor	Search for specified word in all help entries.	159
who	Displays variables currently in the memory.	19, 86
whos	Displays information on variables in the memory.	19, 86

Predefined variables

Variable	Description	Page
ans	Value of last expression.	19
eps	The smallest difference between two numbers.	19
i	$\sqrt{-1}$	19
inf	Infinity.	19
j	Same as i.	19
NaN	Not a number.	19
pi	The number π.	19

Display formats in the Command Window

Command	Description	Page
format bank	Two decimal digits.	13
format compact	Eliminates empty lines.	13
format long	Fixed-point format with 14 decimal digits.	13
format long e	Scientific notation with 15 decimal digits.	13
format long g	Best of 15-digit fixed or floating point.	13
format loose	Adds empty lines.	13
format short	Fixed-point format with 4 decimal digits.	13
format short e	Scientific notation with 4 decimal digits.	13
format short g	Best of 5-digit fixed or floating point.	13

Elementary math functions

Function	Description	Page
abs	Absolute value.	14
exp	Exponential.	14
factorial	The factorial function.	15
log	Natural logarithm.	14
log10	Base 10 logarithm.	14
nthroot	Real nth root or a real number.	14
sqrt	Square root.	14

Trigonometric math functions

Function	Description	Page	Function	Description	Page
acos	Inverse cosine.	15	cos	Cosine.	15
acot	Inverse cotangent.	15	cot	Cotangent.	15
asin	Inverse sine.	15	sin	Sine.	15
atan	Inverse tangent.	15	tan	Tangent.	15

Hyperbolic math functions

Function	Description	Page	Function	Description	Page
cosh	Hyperbolic cosine.	15	sinh	Hyperbolic sine.	15
coth	Hyperbolic cotangent.	15	tanh	Hyperbolic tangent.	15

Rounding

Function	Description	Page
ceil	Round towards infinity.	15
fix	Round towards zero.	15
floor	Round towards minus infinity.	15
rem	Returns the remainder after x is divided by y.	15
round	Round to the nearest integer.	15
sign	Signum function.	16

Creating arrays

Function	Description	Page
diag	Creates a diagonal matrix from a vector. Creates a vector from the diagonal of a matrix.	47
eye	Creates a unit matrix.	38, 63
linspace	Creates equally spaced vector.	36
ones	Creates an array with ones.	38
rand	Creates an array with random numbers.	71, 72
randn	Creates an array with normally distributed numbers.	73
randperm	Creates vector with permutation of integers.	72
zeros	Creates an array with zeros.	38

Handling arrays

Function	Description	Page
length	Number of elements in the vector.	47
reshape	Rearrange a matrix.	47
size	Size of an array.	47

Array functions

Function	Description	Page
cross	Calculates cross product of two vectors.	71
det	Calculates determinant.	64, 71
dot	Calculates scalar product of two vectors.	60, 71
inv	Calculates the inverse of a function.	63, 71
max	Returns maximum value.	70
mean	Calculates mean value.	70
median	Calculates median value.	70
min	Returns minimum value.	70
sort	Arranges elements in ascending order.	70

Array functions (Continued)

Function	Description	Page
std	Calculates standard deviation.	71
sum	Calculates sum of elements.	70

Input and output

Command	Description	Page
disp	Displays output.	91
fprintf	Displays or saves output.	93-100
input	Prompts for user input.	89
load	Retrieves variables to the workspace.	102
save	Saves the variables in the workspace.	101
uiimport	Starts the Import Wizard	106
xlsread	Imports data from Excel	104
xlswrite	Exports data to Excel	105

Two-dimensional plotting

Command	Description	Page
bar	Creates a vertical bar plot.	138
barh	Creates a horizontal bar plot.	138
errorbar	Creates a plot with error bars.	137
fplot	Plots a function.	126
hist	Creates a histogram.	140-142
hold off	Ends hold on.	128
hold on	Keeps current graph open.	128
line	Adds curves to existing plot.	129
loglog	Creates a plot with log scale on both axes.	135
pie	Creates a pie plot.	139
plot	Creates a plot.	120
polar	Creates a polar plot.	142
semilogx	Creates a plot with log scale on the x axis.	135
semilogy	Creates a plot with log scale on the y axis.	135
stairs	Creates a stairs plot.	139
stem	Creates a stem plot.	139

Three-dimensional plotting

Command	Description	Page
bar3	Creates a vertical 3-D bar plot.	275
contour	Creates a 2-D contour plot.	274
contour3	Creates a 3-D contour plot.	274
cylinder	Plots a cylinder.	275
mesh	Creates a mesh plot.	271, 272
meshc	Creates a mesh and a contour plot.	273
meshgrid	Creates a grid for a 3-D plot.	269
meshz	Creates a mesh plot with a curtain.	273
pie3	Creates a pie plot.	276
plot3	Creates a plot.	267
scatter3	Creates a scatter plot.	276
sphere	Plots a sphere.	275
stem3	Creates a stem plot	276
surf	Creates a surface plot.	271, 273

Three-dimensional plotting (Continued)

Command	Description	Page
surfc	Creates a surface and a contour plot.	273
surfl	Creates a surface plot with lighting.	274
waterfall	Creates a mesh plot with a waterfall effect.	274

Formatting plots

Command	Description	Page
axis	Sets limits to axes.	133
colormap	Sets color.	272
grid	Adds grid to a plot.	134, 272
gtext	Adds text a plot.	131
legend	Adds legend to a plot.	131
subplot	Creates multiple plots on one page.	143
text	Adds text a plot.	131
title	Adds title to a plot.	130
view	Controls the viewing direction of a 3-D plot.	277
xlabel	Adds label to x axis.	130
ylabel	Adds label to y axis.	130

Math functions (create, evaluate, solve)

Command	Description	Page
feval	Evaluates the value of a math function.	174
fminbnd	Determines the minimum of a function.	292
fzero	Solves an equation with one variable.	290
inline	Creates an inline function.	169

Numerical integration

Function	Description	Page
quad	Integrates a function.	294
quadl	Integrates a function.	295
trapz	Integrates a function.	296

Ordinary differential equation solvers

Command	Description	Page
ode113	Solves a first order ODE.	298
ode15s	Solves a first order ODE.	299
ode23	Solves a first order ODE.	298
ode23s	Solves a first order ODE.	299
ode23t	Solves a first order ODE.	299
ode23tb	Solves a first order ODE.	299
ode45	Solves a first order ODE.	298

Logical Functions

Function	Description	Page
all	Determines if all array elements are nonzero.	198
and	Logical AND.	197
any	Determines if any array elements are nonzero.	198
find	Finds indices of certain elements of a vector.	198
not	Logical NOT.	197
or	Logical OR.	197

Logical Functions (Continued)

Function	Description	Page
xor	Logical exclusive OR.	198

Flow control commands

Command	Description	Page
break	Terminates execution of a loop.	218
case	Conditionally execute commands.	205
continue	Terminates a pass in a loop.	218
else	Conditionally execute commands.	202
elseif	Conditionally execute commands.	204
end	Terminates conditional statements and loops.	200, 206, 209, 213
for	Repeats execution of a group of commands.	209
if	Conditionally execute commands.	200
otherwise	Conditionally execute commands.	206
switch	Switches among several cases based on expression.	206
while	Repeats execution of a group of commands.	213

Polynomial functions

Function	Description	Page
conv	Multiplies polynomials.	239
deconv	Divides polynomials.	239
poly	Determines coefficients of a polynomial.	238
polyder	Determines the derivative of a polynomial.	240
polyval	Calculates the value of a polynomial.	236
roots	Determines the roots of a polynomial.	237

Curve fitting and interpolation

Function	Description	Page
interp1	One-dimensional interpolation.	250
polyfit	Curve fit polynomial to set of points.	243

Symbolic Math

Function	Description	Page
collect	Collects terms in an expression.	320
diff	Differentiates an equation.	329
double	Converts number from symbolic form to numerical form	318
dsolve	Solves an ordinary differential equation.	333
expand	Expands an expression.	321
ezplot	Plots an expression.	335
factor	Factors to product of lower order polynomials.	321
findsym	Displays the symbolic variables in an expression.	319
int	integrates an expression.	331
pretty	Displays expression in math format.	323
simple	Finds a form of an expression with fewest characters.	323
simplify	Simplifies an expression.	322
solve	Solves a single equation, or a system of equations.	324
subs	Substitutes numbers in an expression.	338
sym	Creates symbolic object.	314
syms	Creates symbolic object.	316

Answers to Selected Problems

Chapter 1

2. *a*) 738.7546   *b*) -0.0732

4. *a*) 0.2846   *b*) 0.1704

6. *a*) 434.1261   *b*) -104.1014

8. *a*) 21.7080 cm   *b*) 24.1799 cm

14. *a*) $\gamma = 126.8699°$
    b) $\alpha = 16.2602°$ $\beta = 36.8699°$

16. $d = 3.2967$

18. *a*) \$266.60
    b) \$281.93
    c) \$282.00

20. *a*) 3.5033   *b*) 3.5769

24. 33 years and 215 day.

26. 0.6325 Pa   17.7828

Chapter 3

2. 0 4.0000 4.4044 4.1603 3.6000
   2.8470 1.9617 0.9788 -0.0801

4. 0 7.8984 12.9328 16.1417 18.1870
   19.4907 20.3217 20.8513 21.1889
   21.4041 21.5413

6. 0 180.7792 323.3089 427.9925
   495.4815 526.8909 524.2756
   491.7770 438.6131 386.7074
   381.6201

8. 4.4861   5.4123   6.1230   6.7221
   7.2500

16.   0   0
    1.0641e+001 3.2680e+003
    2.1281e+001 6.5361e+003
    3.1922e+001 9.8041e+003

4.2562e+001 1.3072e+004
5.3203e+001 1.6340e+004
6.3844e+001 1.9608e+004
7.4484e+001 2.2876e+004
8.5125e+001 2.6144e+004
9.5765e+001 2.9412e+004
1.0641e+002 3.2680e+004

18. $x = 5$, $y = 7$, $z = -2$, $u = 4$,
    $w = 8$

Chapter 4

2. theta = 37.2750   28.1630   22.0930
   18.0097   15.1373

4. 1.0000 0.7937 0.6300 0.5000 0.3969
   0.3150 0.2500 0.1984 0.1575 0.1250
   0.0992 0.0787 0.0625

6.

Time (s)	Distance (m)	Velocity (m/s)
0	0	0
1.0000	0.7750	1.5500
2.0000	3.1000	3.1000
3.0000	6.9750	4.6500
4.0000	12.4000	6.2000
5.0000	19.3750	7.7500
6.0000	27.9000	9.3000
7.0000	37.9750	10.8500
8.0000	49.6000	12.4000
9.0000	62.7750	13.9500
10.0000	77.5000	15.5000

8. The first three rows are:

Temp	SO2	SO3	O2	N2
200	45.5448	63.7192	31.1834	29.1879
220	46.0876	64.9252	31.3675	29.1762
240	46.6101	66.0801	31.5473	29.1584

10.

Resistance (Ohms)	Current (Amps)	Power (Watts)
20.00	2.40	115.20
34.00	1.41	67.76

36.00	1.33	64.00
45.00	1.07	51.20
60.00	0.80	38.40
10.00	4.80	230.40

The source current circuit is 11.811765 Amps.
The total power dissipated in the circuit is 566.964706 Watts.

12. $a = 0.4$, $b = -0.2$, $c = -8$
    $d = -7$, $e = 20$

14. $a = 3$, $b = 8$, $c = 8$, $d = 3$,
    $e = 3$, $f = 8$, $g = 4$

Chapter 5

2.

4.

6.

8.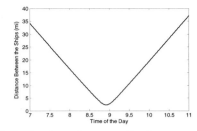

Visibility less than 8 mile from about 8.5 AM until 9.3 AM.

10.

12.

14.

16. 54.39 mm

18. $1.4374e+008$ mm^4

20. *a)*

16.

b)

18.

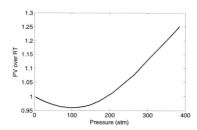

Chapter 7

2. *a)* y = 0
    b) y = 11
    c) y = 0

4. y = [0 -1 -2 3]

6. *a)* New York 37.6774 °F
        Anchorage 33.1290 °F
    b) New York 17
        Anchorage 13
    c) 11 days, on days: 1   7   9
        14   15   18   19   21   22   25
        26
    d) 1 day, on the 23rd.
    e) 16 days, on days: 7   8   9   13
        14   15   16   17   18   19   20
        23   24   25   26   27

Chapter 6

2. *a)* 177.8 cm 79.3651 kg

4. 63.7941 ft/s

6. *a)* (-2.25, -30.125)
    b) (2.5, 68.75)

8. 0.0188 lb

10. *a)*   0.5661   0.7686   0.2979
    b)   -0.2540   -0.8890   -0.3810
    c)   -0.7071   0.7071   0

14. *a)*   82.2833
    b)   83.3667
          61.2667
          71.3500
          82.5333

8.

$$\begin{bmatrix} 0 & -0.333 & -0.5 & -0.6 & -0.6667 \\ 0.3333 & 0 & 0.2 & -0.3333 & -0.4286 \\ 0.5 & 0.2 & 0 & -0.1429 & -0.25 \end{bmatrix}$$

12. 1.0978

20. (11.3099, 15.2971)
    (120.2564, 13.8924)
    (207.8973, 19.2354)
    (-33.0239, 11.9269)

22.

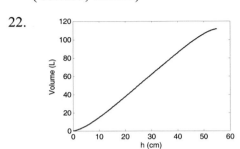

Chapter 8

2. $3x^4 - 2x^2 + 4$

4. x = 0.0022785 m

6. 0.5824 m

8. *a*)   x = 1.1667, y = 9.9167, W = 2
    b)   x = -1.1, y = 21.05, W = 1

10. 1.1987 L

12. *a*)

Pop1975 = 610.0063

b)

Pop1975 = 612.0681

c)

Pop1975L = 614
Pop1975S = 611.8071

14.

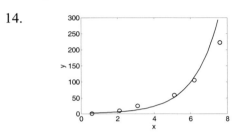

16. m = 2.1666,   b = -1.5928

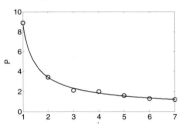

18. R = 0.08215682326924
    (Units of R: L-atm/mol-K)

Chapter 9

2.

4.

6.

8.

10. *a*)

b)

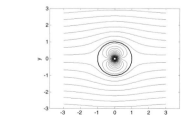

$$1/(2\pi\sqrt{LC}) = 83.882\,\text{Hz}$$

12.

14.

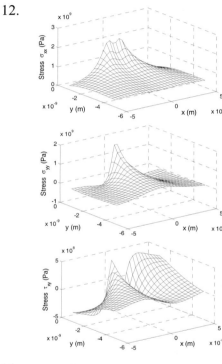

Chapter 10

2. 1.3923, 2.0714, 3.1895

4. 7.2792 m

6.   0.5405 V

8. R_1 = 3.9026 cm, h = 4.18 cm.

10. h = 22.6667 cm, r = 16.0278 , cm

12. 3.5933

14. E = 6.0986e+006 N/C

16. 61275 square miles

18. 3.6531e+010 km, 1.6815e+004 km/h

20.

22.

24.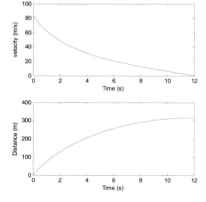

Chapter 11

2.   *a)*   27-x^3
     b)   (3-x)/(9+3*x+x^2)

c)   12+x^2+2*x
d)   36

4.   *a)*   (x-1)*(x-2)*(x-5)*(x+4)*(x+3)
     b)   x^4-4*x^3-19*x^2+46*x+120

10. *a)*   T = 5/12*W*x   FAx =200/3*x
          FAy =-50*x+200
    b)   T =250/3*x
    c)

14.   1/9*sin(x)-4/27*log(2+3*sin(x))-4/27/(2+3*sin(x))

20.   x = exp(-R*(N+1)*t)*N*(N+1)/(1+exp(-R*(N+1)*t)*N)
       t_max = log(N)/R/(N+1)

22. *a)*   g/c*m-exp(-c/m*t)*g/c*m
    b)   16.1489 kg/s
    c)

24.   ys = -11/50*cos(5^(1/2)*t)+11/50-1/20*t^2

Index